Forest and Plant Bioproducts Division 2015

Core Programming Area at the 2015 AIChE Annual Meeting

Salt Lake City, Utah, USA
8-13 November 2015

ISBN: 978-1-5108-1861-3

Printed from e-media with permission by:

Curran Associates, Inc.
57 Morehouse Lane
Red Hook, NY 12571

Some format issues inherent in the e-media version may also appear in this print version.

Printed by Curran Associates, Inc. (2016)

For permission requests, please contact AIChE
at the address below.

AIChE
120 Wall Street, FL 23
New York, NY 10005-4020

Phone: (800) 242-4363
Fax: (203) 775-5177

www.aiche.org

Additional copies of this publication are available from:

Curran Associates, Inc.
57 Morehouse Lane
Red Hook, NY 12571 USA
Phone: 845-758-0400
Fax: 845-758-2634
Email: curran@proceedings.com
Web: www.proceedings.com

TABLE OF CONTENTS

437193 Overview of Cellulose Nanomaterial Applications

Abstract:

Nanocellulose has proven to be a versatile material with a vast array of potential commercial applications including composite, automotive, aerospace, and building construction, modifiers for cosmetics and oil drilling fluids, and high performance fillers for paper, packaging, paints, plastics, biomedical engineering, electronics, and cement. In addition to material performance properties like gelation, shear thinning, exceptionally high strength, and light weight, nanocellulose has a strong sustainability profile. Being made from biomass, it is renewable, biodegradable, compostable, and designed for the environment with a sustainable life cycle carbon footprint.

This talk will give a brief review on our recent research in the applications of cellulose based nanomaterials, including chemical surface modification of nanocellulose materials, barrier package, aerogel, composites, and electronic applications.

437194 Emulsion of Poly (lignin-co-butyl acrylate) s As Biobased Coating Films

Abstract:

Stable waterborne latex are gaining a lot of importance due to its environment-friendly production, ease in manipulation, and a variety of application areas including packaging, paints and coatings, polymer foams, and films. For instance, latex usually includes rigid styrene and expandable acrylic monomers, which provide proper film formation properties and physical strength through waterborne synthesis methodologies. In order to increase sustainability of current latex film, this work aims to use lignin to substitute petroleum-based styrene to design a waterborne emulsion for coating and film formation materials with butylacrylate monomer together. The lignin was modified to contain polymerizable acrylic groups, which was acted as macromonomers for free radical polymerization with butylacrylate together via emulsion polymerization. The emulsion was then cast to form films. The ^{1}H-NMR and P^{31}-NMR confirmed the synthesis of lignin-based acrylate. The lignin-based copolymer films were characterized by thermogravimetric analysis (TGA), tensile testing, and gel permeation chromatograph (GPC) and scanning electron microscope (SEM). The results showed that the increase in modified lignin monomer (even in low concentrations) in the emulsion system improved the mechanical and thermal properties of films formed. The SEM images of the copolymer film indicated the growth of the polybutyl acrylate particles along lignin cores as templates. In summary, lignin can be a feasible candidate for sustainable and green acrylate latex film synthesis.

437196 Potential Applications of Functionalized Nanocrystalline Cellulose

Abstract:

NCC represents a new generation of renewable nano-biomaterials with unique physical, chemical and optical properties with vast potential applications. The unique properties of NCC have afforded its promising applications in many industrial sectors. However despite significant advances toward large scale production of nanocrystalline cellulose from forest biomass, there has been little success toward identifying a marketable product justifying the manufacture cost of NCC. In the presentation, we will discuss recent work on functionalization of nanocrystalline cellulose for potential high value applications.

437048 UV Resistibility of Polystyrene Co-Butyl Acrylate (PSBA) Encapsulated Lignin-Saponite Nanohybrid Composite Film

Abstract:

Protecting biological systems and organic materials from high ultraviolet (UV) radiation is a significant issue for engineering materials in both terrestrial and space applications. In this study, the synthesized composites of polystyrene butylacrylate (PSBA) encapsulated lignin based nanohybrid were evaluated as UV resistant cladding materials for both applications. The UV transmittance of the composite films and their physical, thermal, and chemical degradation after prolonged exposure to high UV radiation (10 Earth days and 30 Mars equivalent days) at low temperatures of 3.4 °C were examined. The results showed that the composite films blocked the UV radiation and there was no significant effect of UV radiation on film integrity, mechanical and thermal properties of the composite films when they were exposed to high intensity UV irradiation (1.3510^7 (KJ/m^2)/day UV flux) within 24h. Further, FTIR spectra of the composite films before and after UV radiation demonstrated that the films didn't have any notable oxidation degradation, although the glass transition temperature was slightly decreased when increasing exposure time.

Biomass Pretreatment using Ionic Liquid-Glycerol Mixtures

Joan G. Lynam* and Charles J. Coronella
AIChE Annual Meeting 2015, Monday, November 9, 2015

Chemical & Materials Engineering Dept.
University of Nevada, Reno

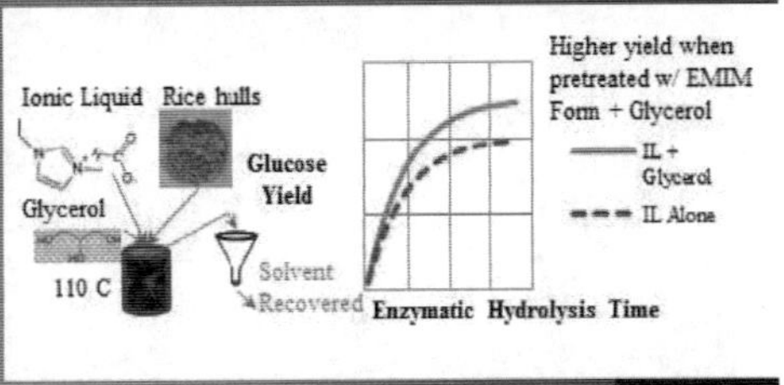

*Presenting

Lignocellulosic Structure

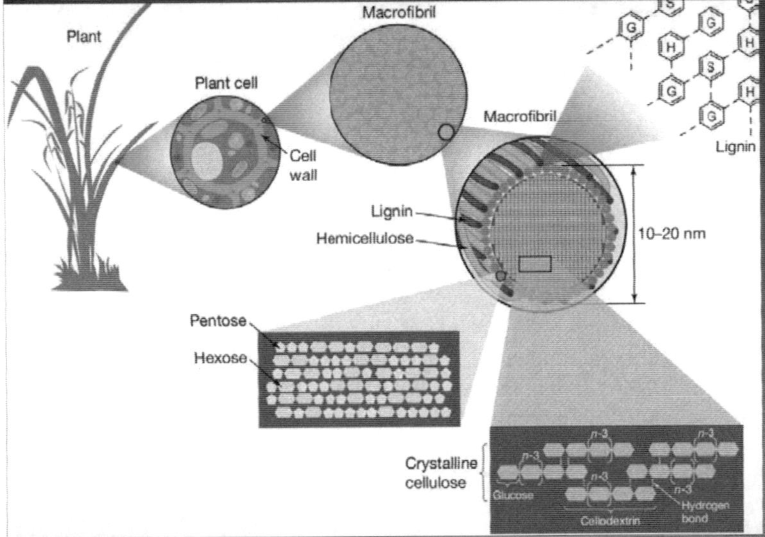

p-hydroxyphenyl (H) guaiacyl (G), and syringyl (S)

Lignin shields cellulose from use.

Hemicelluloses are easiest to remove.

Mannose and Xylose, hemicellulosic monomers of galactoglucomannan and arabinoglucuronoxylan

https://www.detip.upc.edu/recerca/fotos-projectes/projecte1

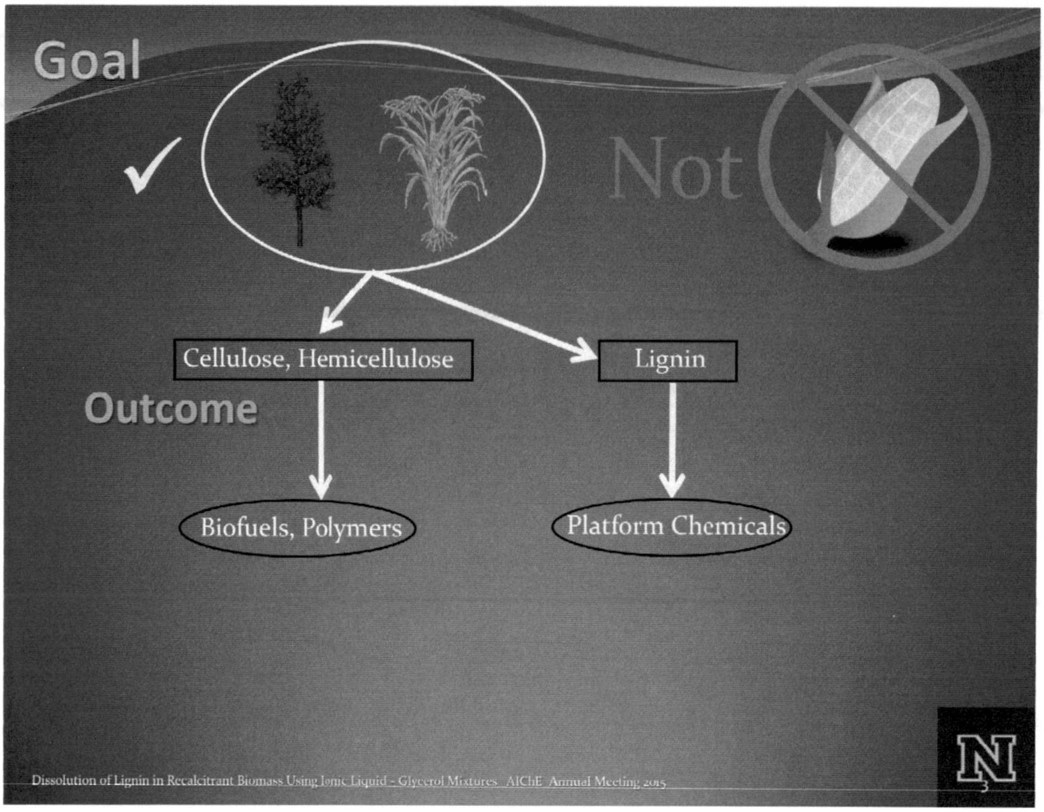
Goal
Not
Cellulose, Hemicellulose
Lignin
Outcome
Biofuels, Polymers
Platform Chemicals
Dissolution of Lignin in Recalcitrant Biomass Using Ionic Liquid - Glycerol Mixtures AIChE Annual Meeting 2015
3

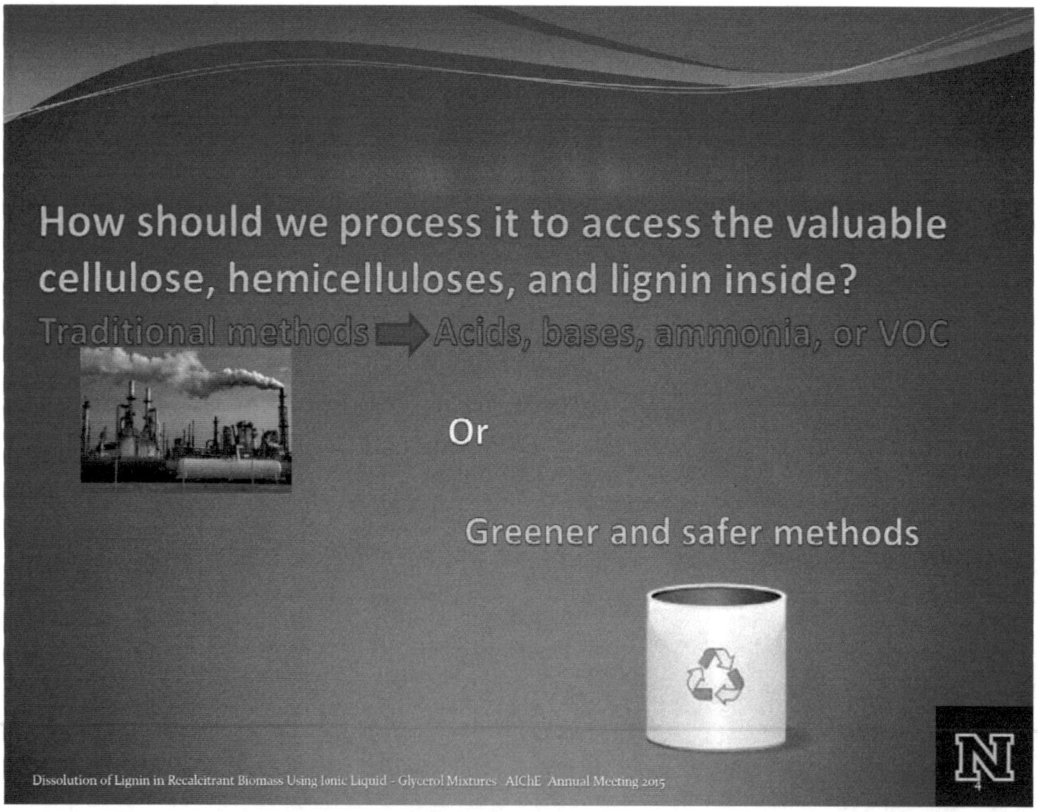
How should we process it to access the valuable cellulose, hemicelluloses, and lignin inside?
Traditional methods
Acids, bases, ammonia, or VOC
Or
Greener and safer methods
Dissolution of Lignin in Recalcitrant Biomass Using Ionic Liquid - Glycerol Mixtures AIChE Annual Meeting 2015
4

Ionic Liquids

1. Molten salts – melting point < 100°C
2. Large cation – ionic structure poorly coordinated
3. Extremely low vapor pressure (unlike VOCs)→"green"
4. Can be recycled
5. Can dissolve lignin in lignocellulosic biomass!

Dissolution of Lignin in Recalcitrant Biomass Using Ionic Liquid - Glycerol Mixtures AIChE Annual Meeting 2015 5

Used these ILs Effective for Biomass

1-Ethyl-3-methylimidazolium acetate (EMIM Ac)

1-Ethyl-3-methylimidazolium formate (EMIM Form)

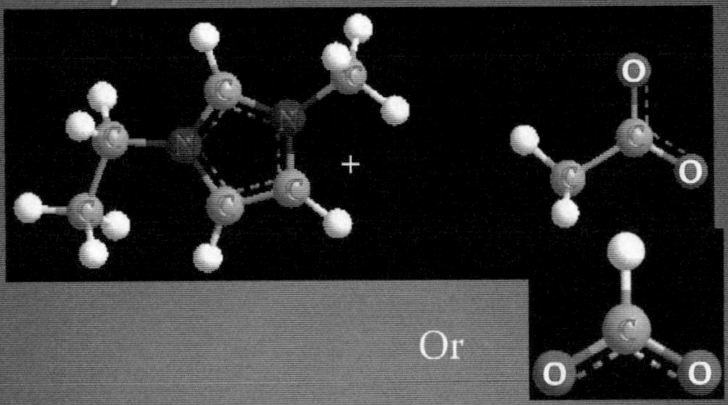

Dissolution of Lignin in Recalcitrant Biomass Using Ionic Liquid - Glycerol Mixtures AIChE Annual Meeting 2015 6

These have high Kamlet-Taft ß values

ß => ability to be hydrogen bond acceptor

1-Ethyl-3-methylimidazolium acetate **1.2** (EMIM Ac)

1-Ethyl-3-methylimidazolium formate **1.0** (EMIM Form)

Kamlet-Taft ß values greater than 0.8 can dissolve cellulose

Used these ILs Effective for Biomass

1-Ethyl-3-methylimidazolium acetate (EMIM Ac)

1-Ethyl-3-methylimidazolium formate (EMIM Form)

But they are costly! Or

$$$$

A low-cost cosolvent is needed to make processing biomass with ILs practical.

Water is an anti-solvent when used with biomass.

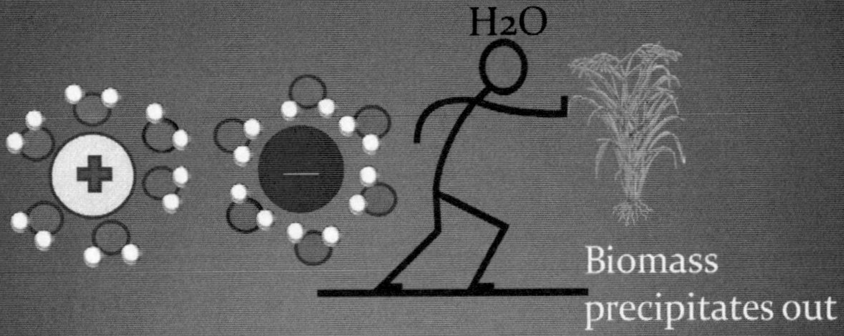

Biodiesel Reaction

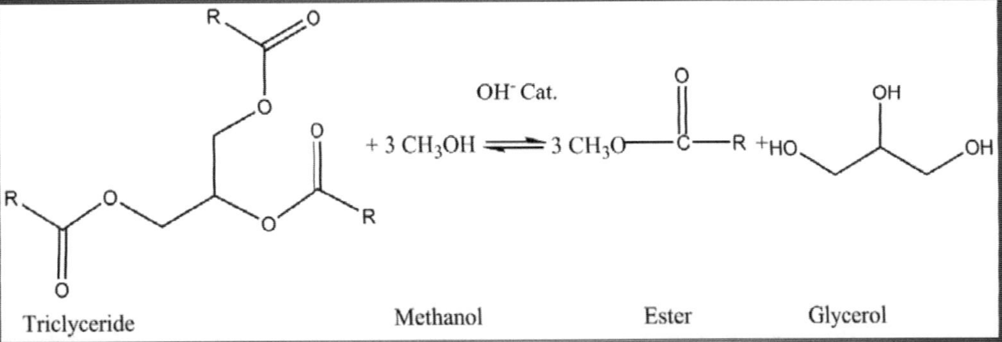

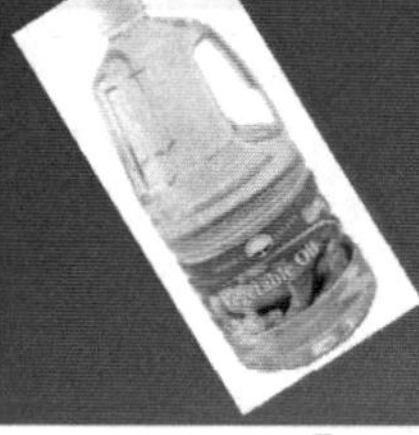

Biodiesel Reaction

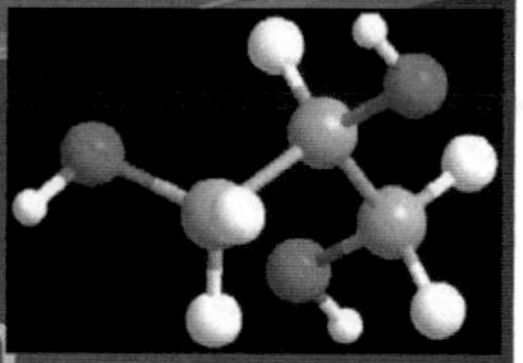

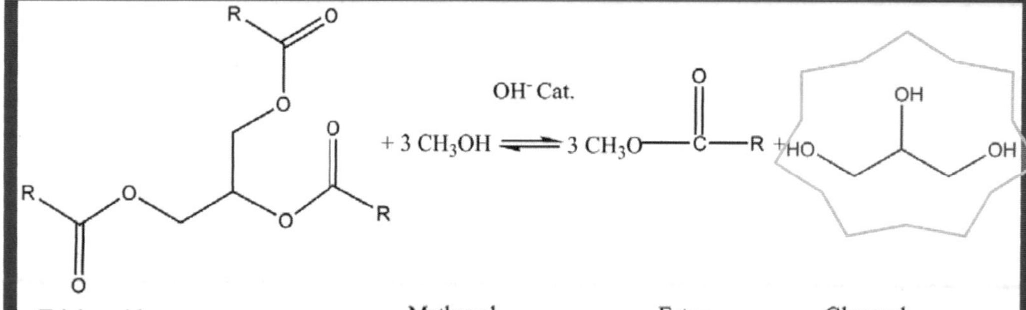

Glycerol is a byproduct of biodiesel production and is about 18¢/kg (8¢/lb).

Biomass Pretreatment Procedure

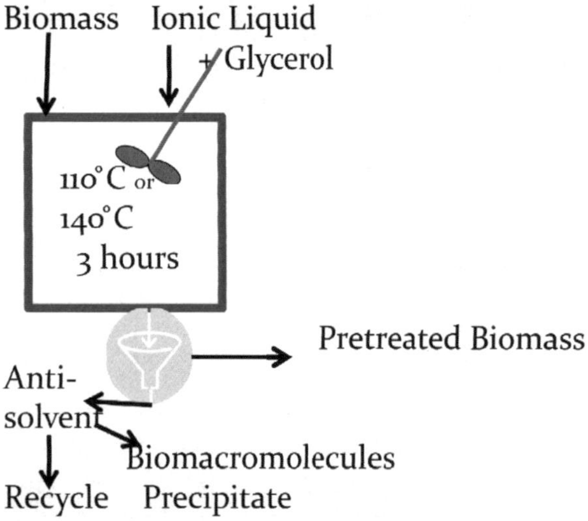

Biomass Pretreatment Procedure

Pretreated biomass rinsed, filtered and dried.

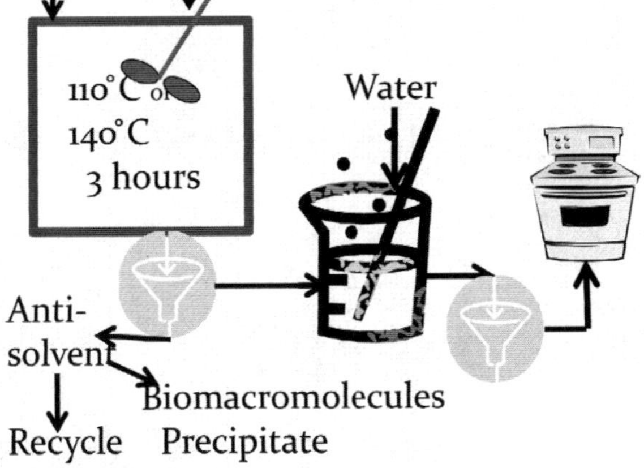

Biomass Pretreatment Procedure

Biomass is enzymatically hydrolyzed and HPLC used to determine glucose and hemicellulosic sugars liberated from the biomass.

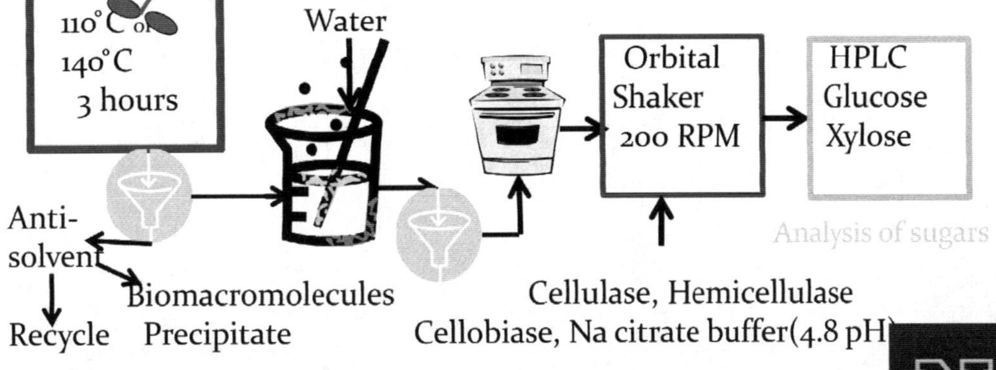

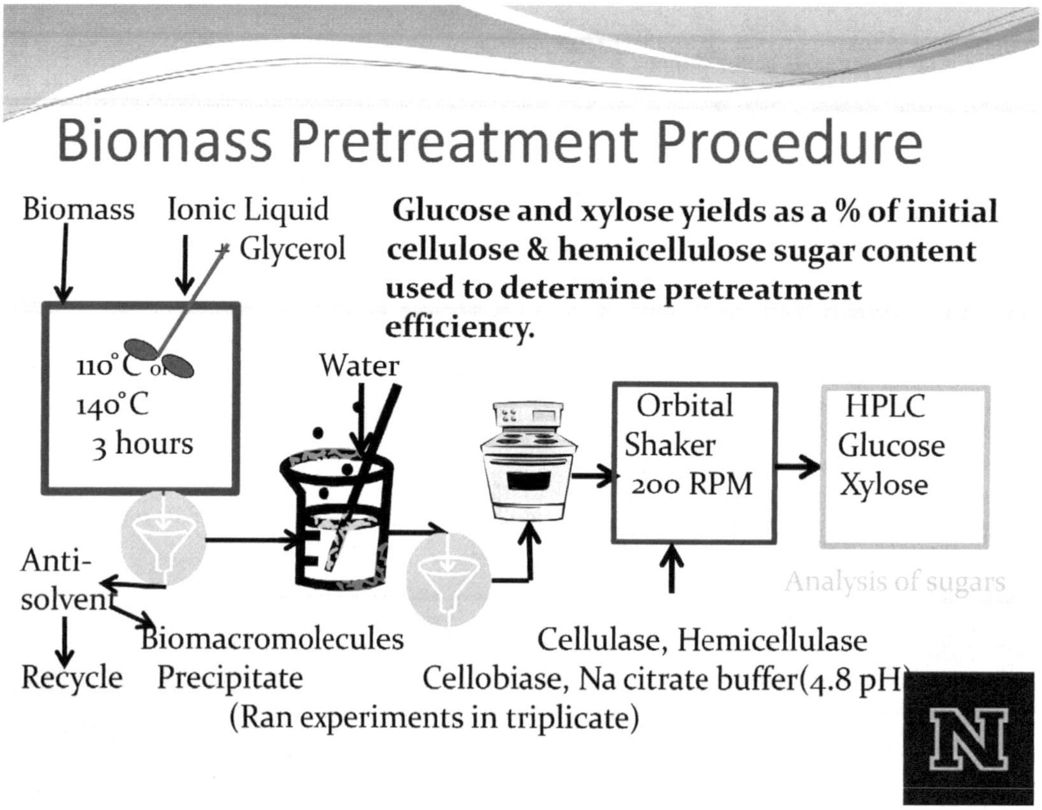
Biomass Pretreatment Procedure
Biomass
Ionic Liquid
+ Glycerol
Glucose and xylose yields as a % of initial cellulose & hemicellulose sugar content used to determine pretreatment efficiency.
110°C or
140°C
3 hours
Water
Orbital
Shaker
200 RPM
HPLC
Glucose
Xylose
Analysis of sugars
Anti-
solvent
Recycle
Biomacromolecules
Precipitate
Cellulase, Hemicellulase
Cellobiase, Na citrate buffer(4.8 pH)
(Ran experiments in triplicate)

Secondary Agricultural Residues – Benefits
Must be generated to produce food
Already removed from field to processing centers
Rice hulls
1200
800
400
0
World Production
Million Tonnes/year
Maize
Rice
Wheat
% Component
Lignin (12%)
Cellulose (39%)
Hemicellulose(16%)
Aqueous solubles (12%)
Ash(~21%)
Faostat 2013
Rice hulls are ~20% of the seed
Dissolution of Lignin in Recalcitrant Biomass Using Ionic Liquid - Glycerol Mixtures AIChE Annual Meeting 2015
16

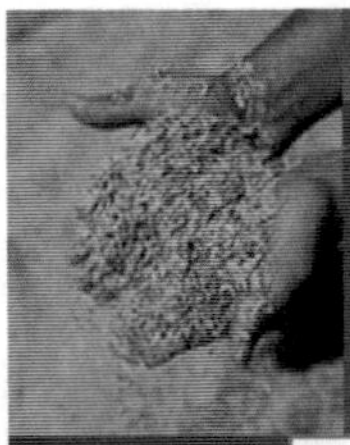

Glucose yield with time

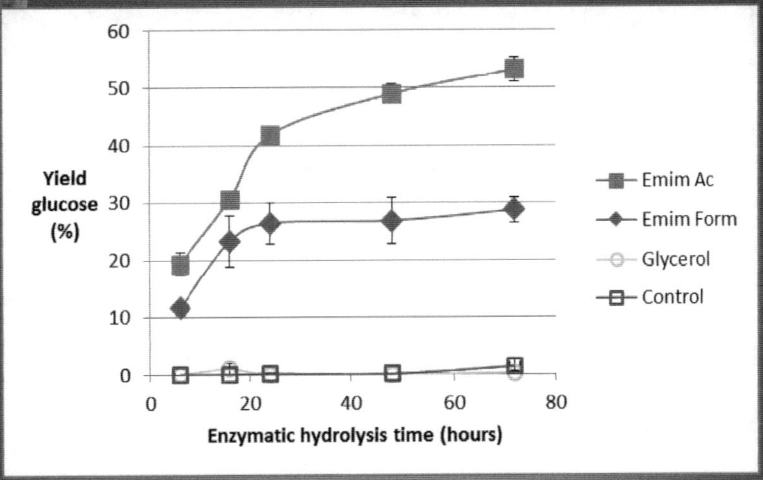

Glucose yield as a percentage of cellulose in original biomass

Control is raw rice hulls

Dissolution of Lignin in Recalcitrant Biomass Using Ionic Liquid - Glycerol Mixtures AIChE Annual Meeting 2015

17

Why is EMIM Ac so much better? It has a higher Kamlet-Taft ß value (1.2) than EMIM Form (1.0)

ß => ability to be hydrogen bond acceptor

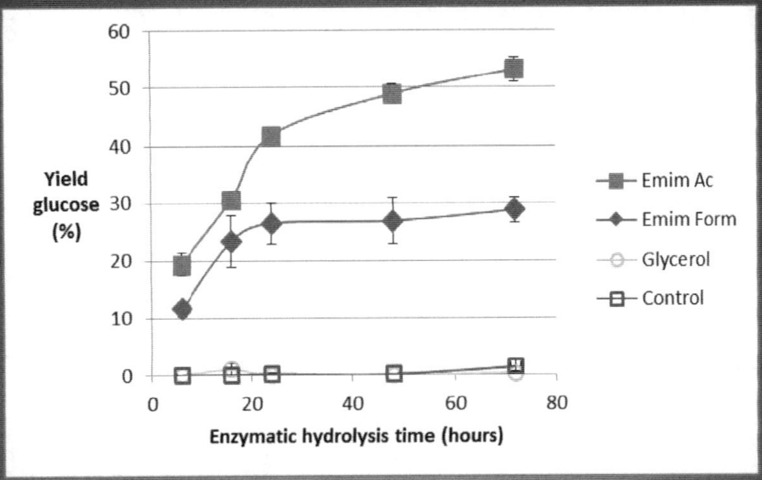

Control is raw rice hulls

Also, EMIM Form has a higher lattice potential energy, so it is more tightly bonded to itself.

Dissolution of Lignin in Recalcitrant Biomass Using Ionic Liquid - Glycerol Mixtures AIChE Annual Meeting 2015

18

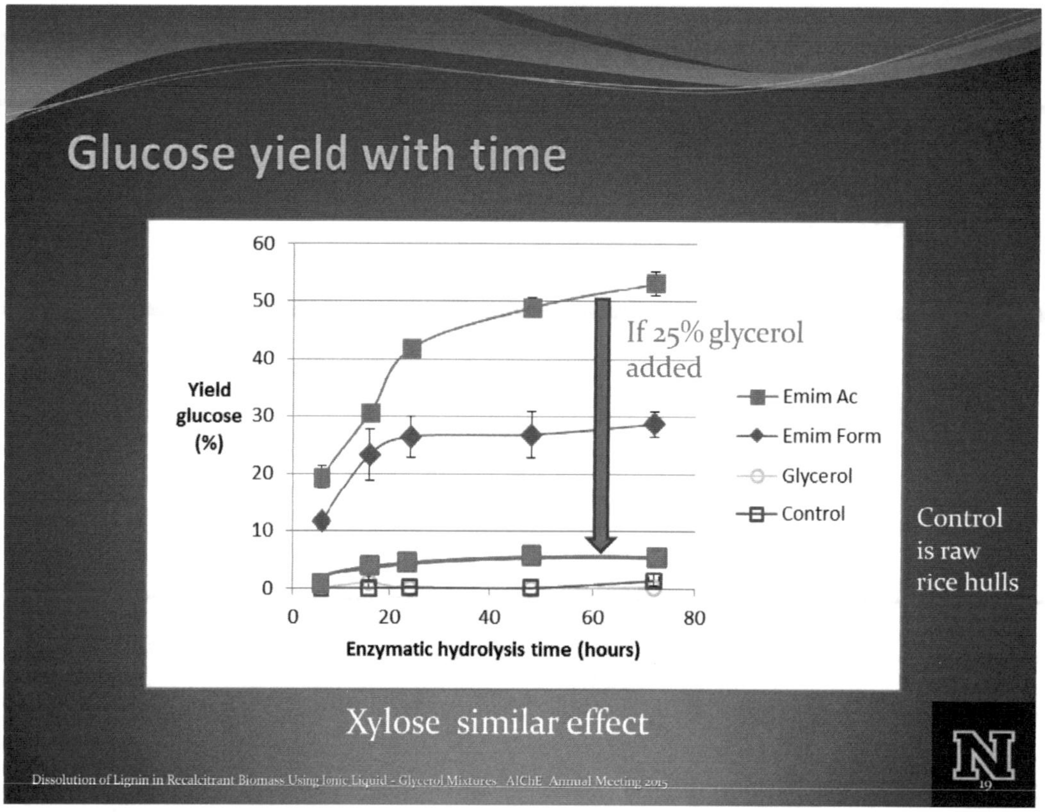

Glucose yield with time
Yield glucose (%)
60
50
40
30
20
10
0
0
20
40
60
80
Enzymatic hydrolysis time (hours)
If 25% glycerol added
Emim Ac
Emim Form
Glycerol
Control
Control is raw rice hulls
Xylose similar effect
Dissolution of Lignin in Recalcitrant Biomass Using Ionic Liquid - Glycerol Mixtures AIChE Annual Meeting 2015
19

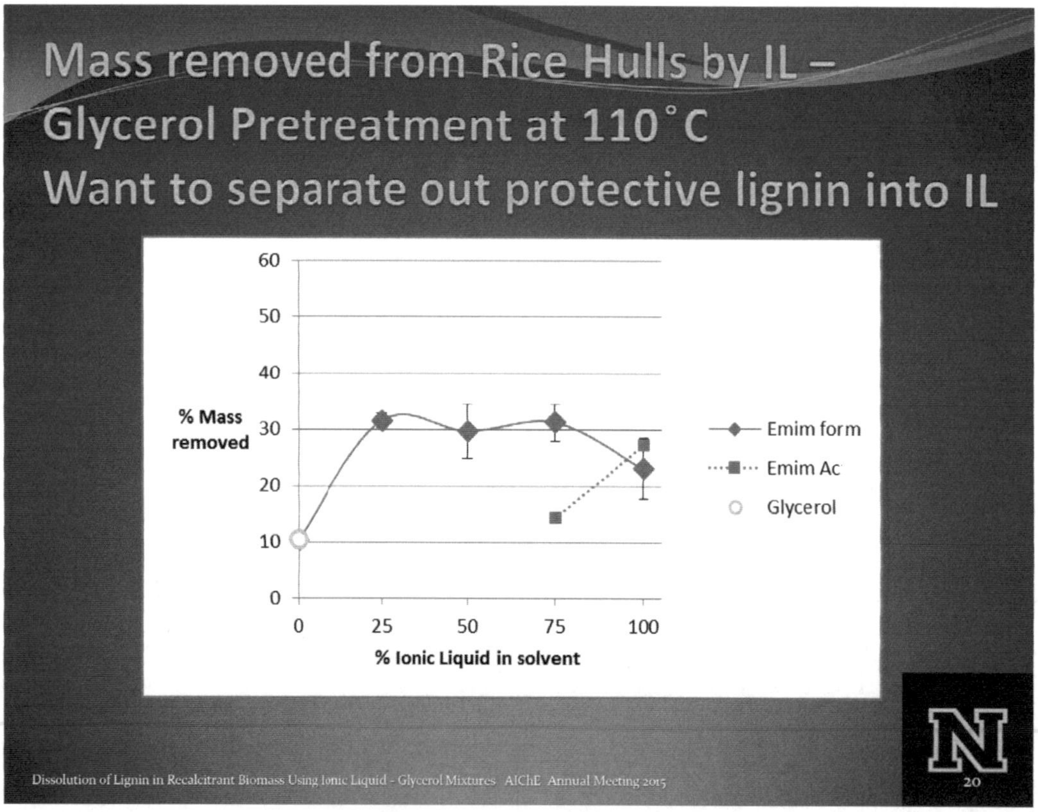

Mass removed from Rice Hulls by IL – Glycerol Pretreatment at 110 °C
Want to separate out protective lignin into IL
% Mass removed
60
50
40
30
20
10
0
0
25
50
75
100
% Ionic Liquid in solvent
Emim form
Emim Ac
Glycerol
Dissolution of Lignin in Recalcitrant Biomass Using Ionic Liquid - Glycerol Mixtures AIChE Annual Meeting 2015
20

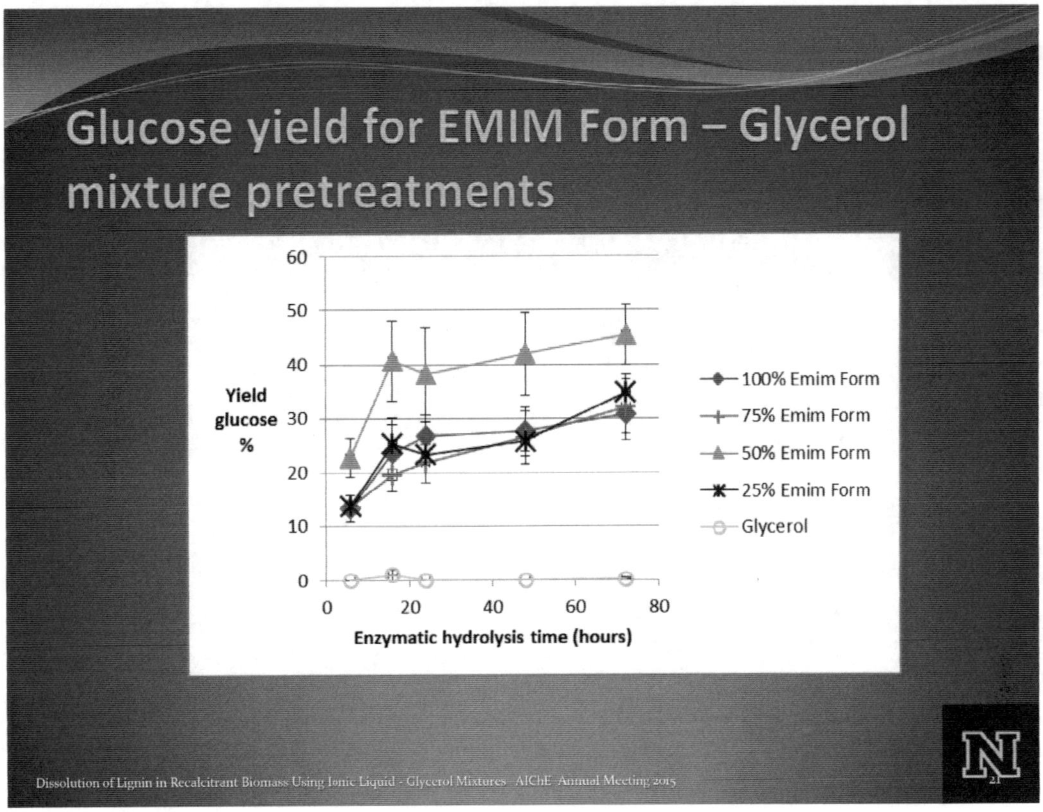

Glucose yield for EMIM Form – Glycerol mixture pretreatments
Yield glucose %
Enzymatic hydrolysis time (hours)
100% Emim Form
75% Emim Form
50% Emim Form
25% Emim Form
Glycerol
Dissolution of Lignin in Recalcitrant Biomass Using Ionic Liquid - Glycerol Mixtures AIChE Annual Meeting 2015
21

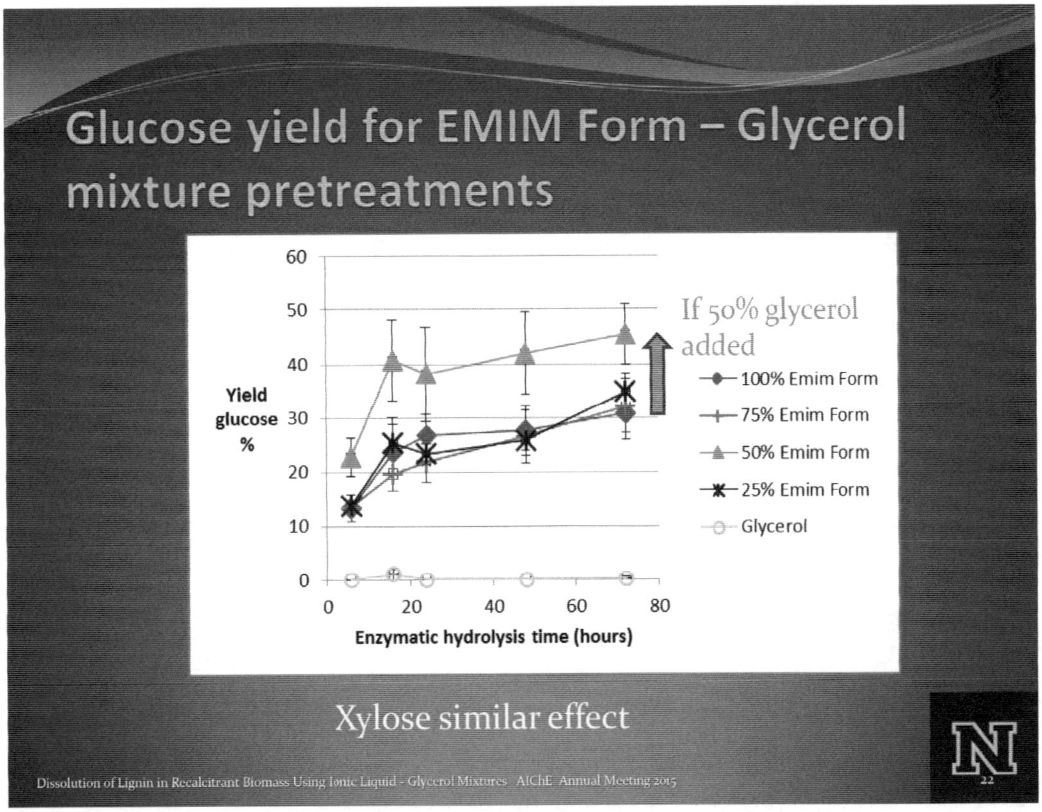

Glucose yield for EMIM Form – Glycerol mixture pretreatments
Yield glucose %
Enzymatic hydrolysis time (hours)
If 50% glycerol added
100% Emim Form
75% Emim Form
50% Emim Form
25% Emim Form
Glycerol
Xylose similar effect
Dissolution of Lignin in Recalcitrant Biomass Using Ionic Liquid - Glycerol Mixtures AIChE Annual Meeting 2015
22

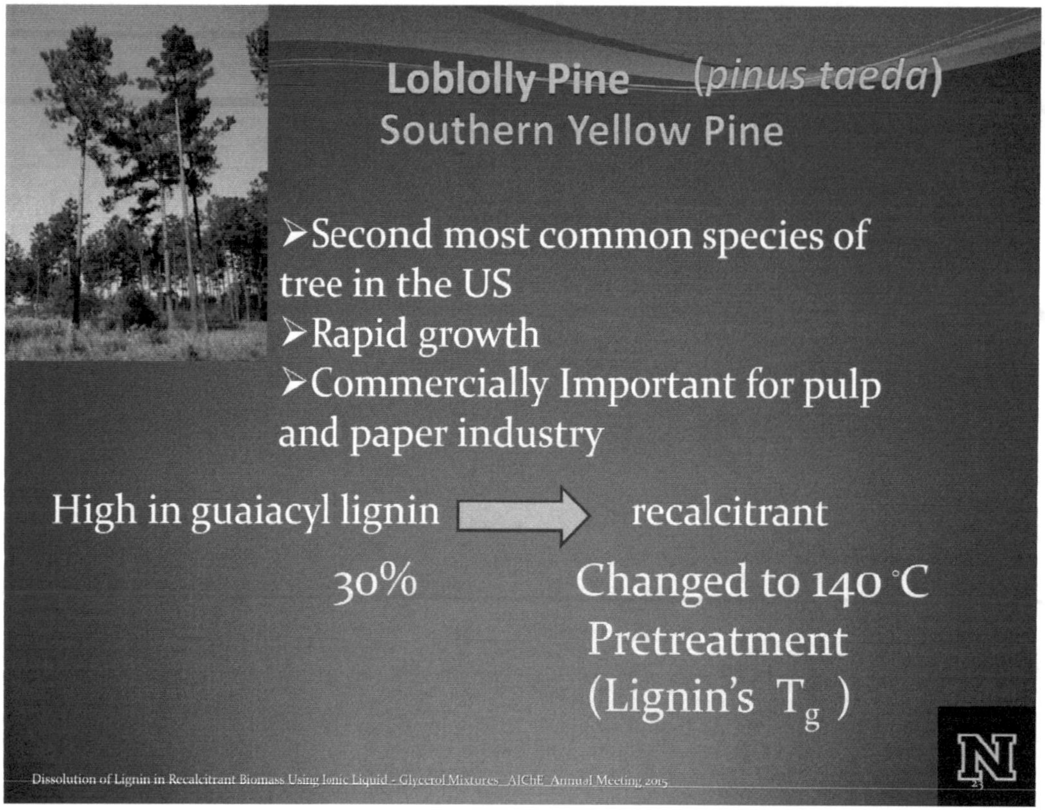
Loblolly Pine (pinus taeda)
Southern Yellow Pine
➢Second most common species of tree in the US
➢Rapid growth
➢Commercially Important for pulp and paper industry
High in guaiacyl lignin
recalcitrant
30%
Changed to 140 °C Pretreatment
(Lignin's T_g)
Dissolution of Lignin in Recalcitrant Biomass Using Ionic Liquid - Glycerol Mixtures AIChE Annual Meeting 2015
23

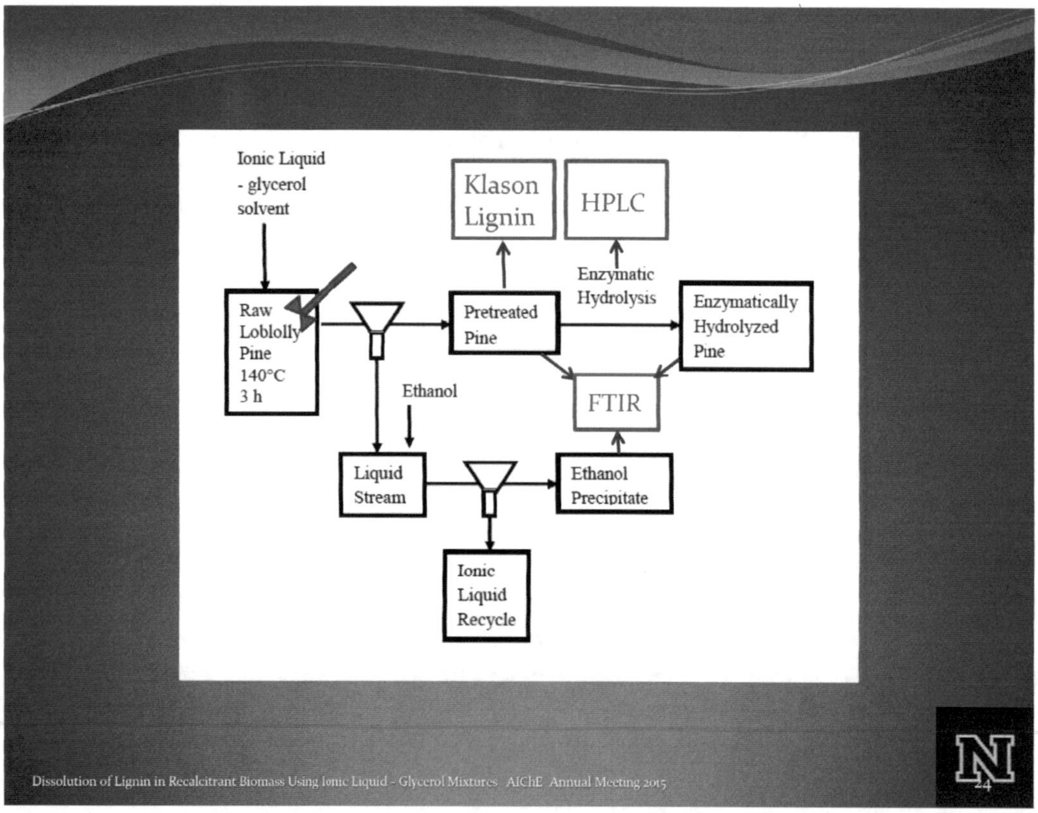
Ionic Liquid - glycerol solvent
Klason Lignin
HPLC
Enzymatic Hydrolysis
Raw Loblolly Pine 140°C 3 h
Pretreated Pine
Enzymatically Hydrolyzed Pine
Ethanol
FTIR
Liquid Stream
Ethanol Precipitate
Ionic Liquid Recycle
Dissolution of Lignin in Recalcitrant Biomass Using Ionic Liquid - Glycerol Mixtures AIChE Annual Meeting 2015
24

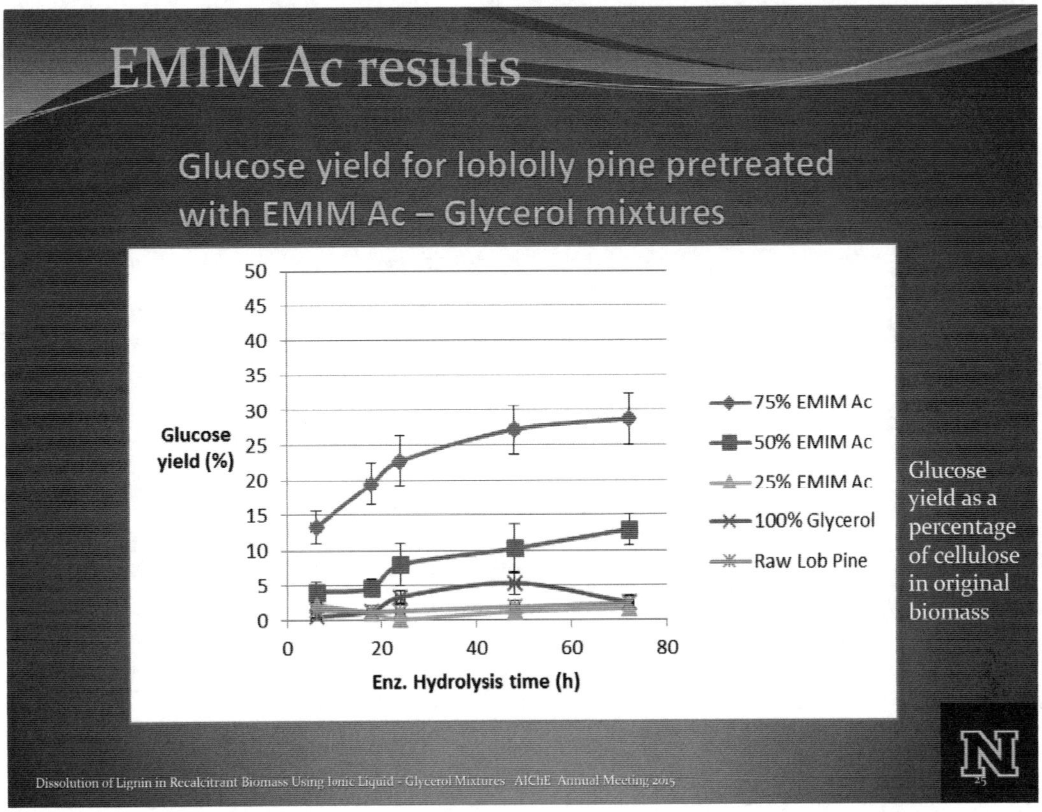
EMIM Ac results
Glucose yield for loblolly pine pretreated with EMIM Ac – Glycerol mixtures
Glucose yield (%)
50
45
40
35
30
25
20
15
10
5
0
0
20
40
60
80
Enz. Hydrolysis time (h)
75% EMIM Ac
50% EMIM Ac
25% EMIM Ac
100% Glycerol
Raw Lob Pine
Glucose yield as a percentage of cellulose in original biomass
Dissolution of Lignin in Recalcitrant Biomass Using Ionic Liquid - Glycerol Mixtures AIChE Annual Meeting 2015
25

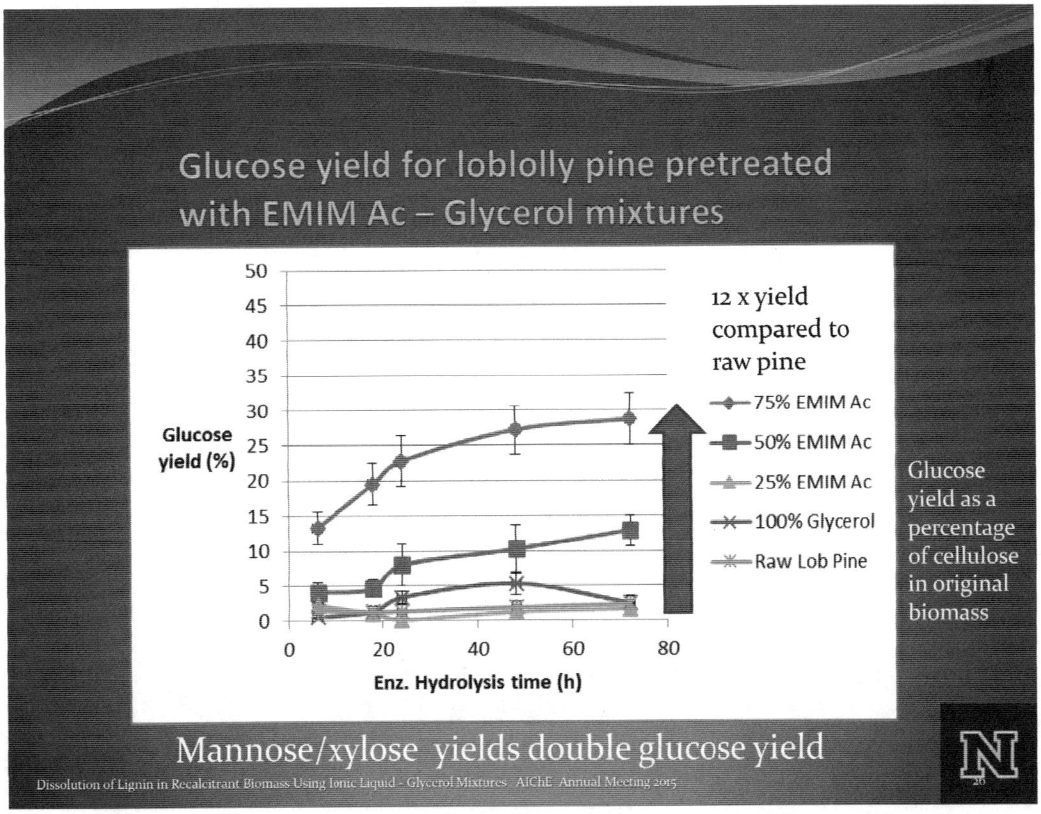
Glucose yield for loblolly pine pretreated with EMIM Ac – Glycerol mixtures
Glucose yield (%)
50
45
40
35
30
25
20
15
10
5
0
0
20
40
60
80
Enz. Hydrolysis time (h)
12 x yield compared to raw pine
75% EMIM Ac
50% EMIM Ac
25% EMIM Ac
100% Glycerol
Raw Lob Pine
Glucose yield as a percentage of cellulose in original biomass
Mannose/xylose yields double glucose yield
Dissolution of Lignin in Recalcitrant Biomass Using Ionic Liquid - Glycerol Mixtures AIChE Annual Meeting 2015
26

EMIM Form results

Glucose yield for loblolly pine pretreated with EMIM Form – Glycerol mixtures

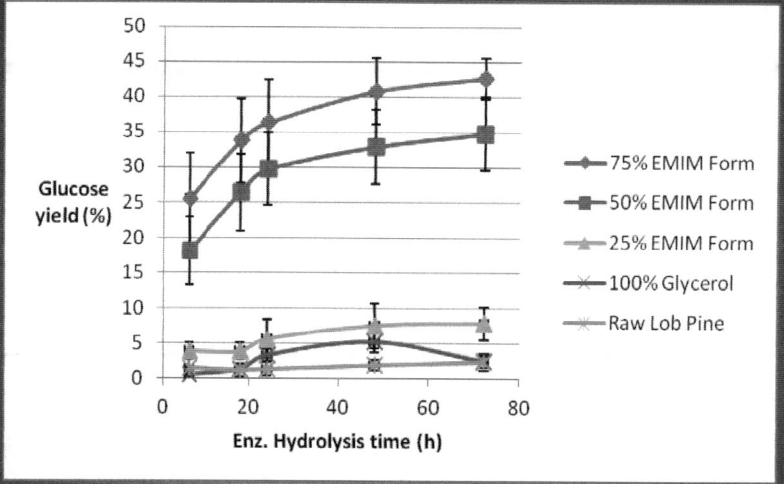

Glucose yield as a percentage of cellulose in original biomass

Dissolution of Lignin in Recalcitrant Biomass Using Ionic Liquid - Glycerol Mixtures AIChE Annual Meeting 2015 27

Glucose yield for loblolly pine pretreated with EMIM Form – Glycerol mixtures

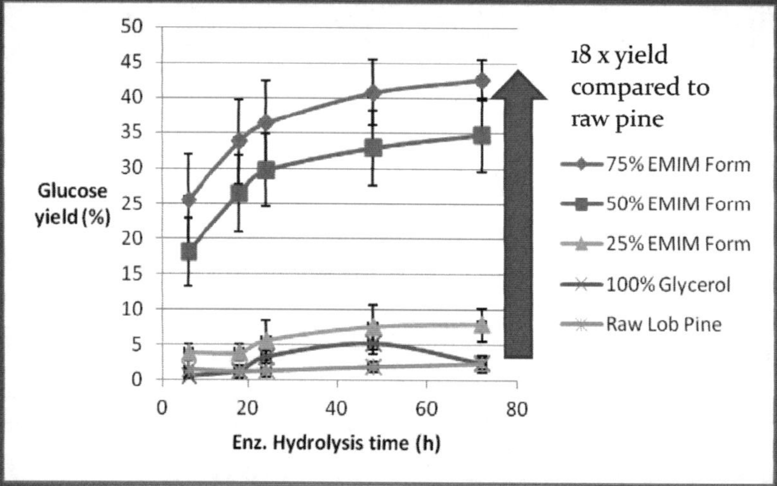

Glucose yield as a percentage of cellulose in original biomass

Mannose/xylose yields double glucose yield for 50% EMIM Form

Dissolution of Lignin in Recalcitrant Biomass Using Ionic Liquid - Glycerol Mixtures AIChE Annual Meeting 2015 28

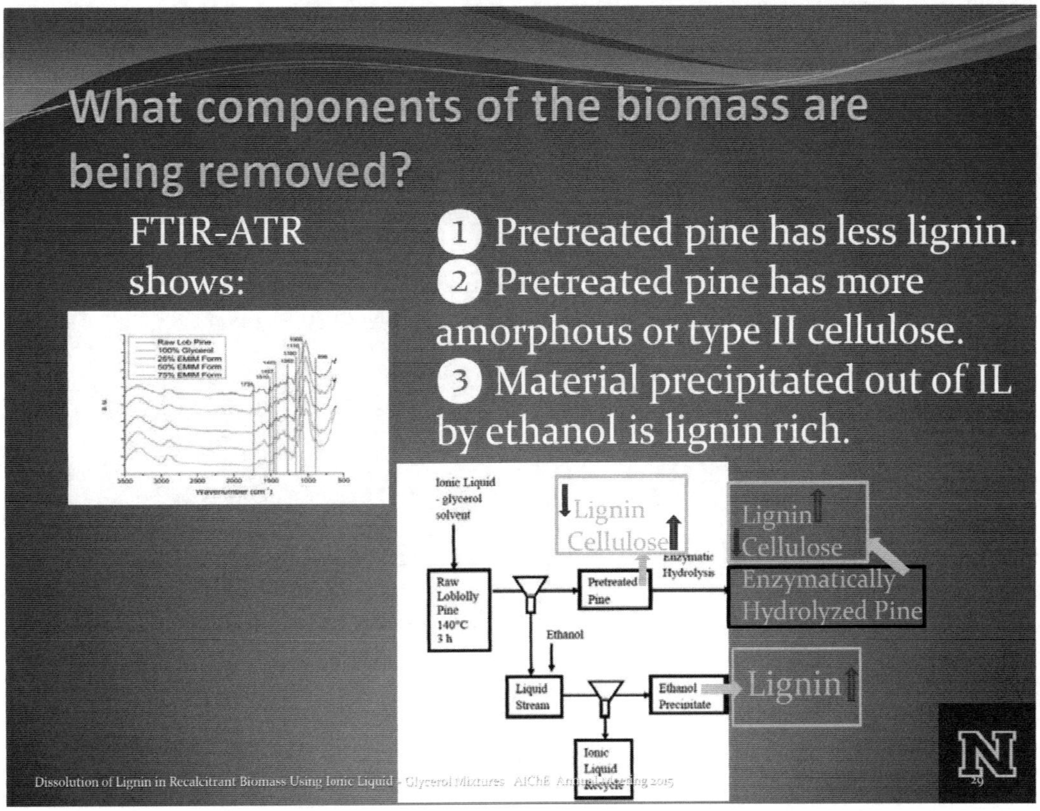
What components of the biomass are being removed?
FTIR-ATR shows:
1 Pretreated pine has less lignin.
2 Pretreated pine has more amorphous or type II cellulose.
3 Material precipitated out of IL by ethanol is lignin rich.
Ionic Liquid - glycerol solvent
Raw Loblolly Pine 140°C 3 h
Pretreated Pine
Enzymatic Hydrolysis
Lignin
Cellulose
Lignin
Cellulose
Enzymatically Hydrolyzed Pine
Ethanol
Liquid Stream
Ethanol Precipitate
Lignin
Ionic Liquid Recycle
Dissolution of Lignin in Recalcitrant Biomass Using Ionic Liquid - Glycerol Mixtures AIChE Annual Meeting 2015
29

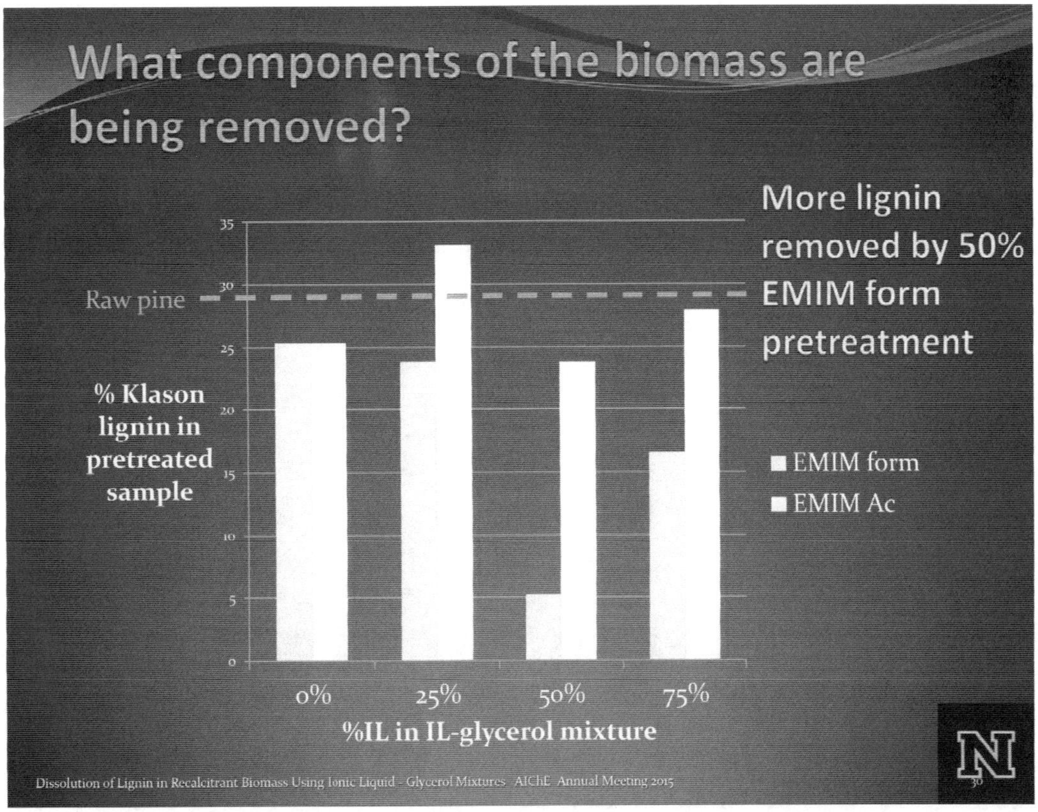
What components of the biomass are being removed?
More lignin removed by 50% EMIM form pretreatment
Raw pine
% Klason lignin in pretreated sample
0
5
10
15
20
25
30
35
0%
25%
50%
75%
%IL in IL-glycerol mixture
EMIM form
EMIM Ac
Dissolution of Lignin in Recalcitrant Biomass Using Ionic Liquid - Glycerol Mixtures AIChE Annual Meeting 2015
30

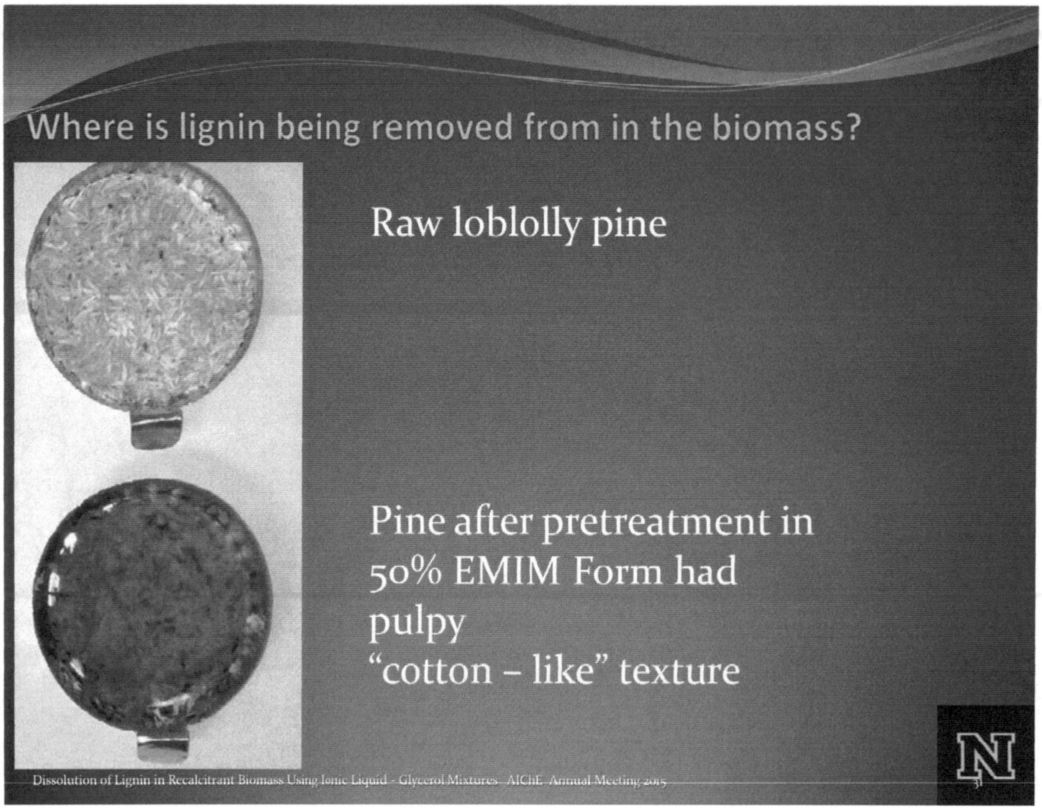
Where is lignin being removed from in the biomass?
Raw loblolly pine
Pine after pretreatment in 50% EMIM Form had pulpy "cotton – like" texture
Dissolution of Lignin in Recalcitrant Biomass Using Ionic Liquid - Glycerol Mixtures AIChE Annual Meeting 2015
31

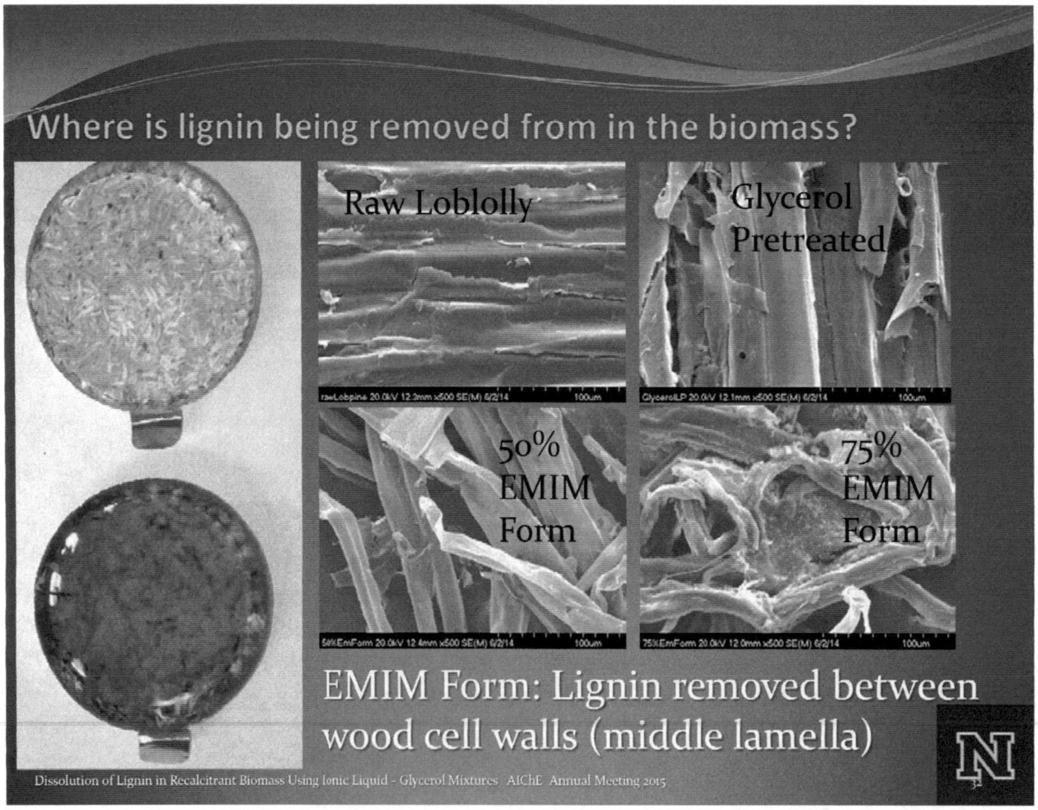
Where is lignin being removed from in the biomass?
Raw Loblolly
Glycerol Pretreated
50% EMIM Form
75% EMIM Form
EMIM Form: Lignin removed between wood cell walls (middle lamella)
Dissolution of Lignin in Recalcitrant Biomass Using Ionic Liquid - Glycerol Mixtures AIChE Annual Meeting 2015
32

Why?

Dissolution of Lignin in Recalcitrant Biomass Using Ionic Liquid - Glycerol Mixtures AIChE Annual Meeting 2015

33

Why?

Why does EMIM Form with glycerol pretreatment work better than EMIM Ac with glycerol?

Why does EMIM Form's performance improve with glycerol dilution?

Dissolution of Lignin in Recalcitrant Biomass Using Ionic Liquid - Glycerol Mixtures AIChE Annual Meeting 2015

34

Why?

Why does EMIM Form with glycerol pretreatment work better than EMIM Acet with glycerol?

Why does EMIM Form's performance improve with glycerol dilution?

To figure it out I investigated pH, viscosity, and density.

Dissolution of Lignin in Recalcitrant Biomass Using Ionic Liquid - Glycerol Mixtures AIChE Annual Meeting 2015 35

pH

1:9 mass ratio solution of IL: water

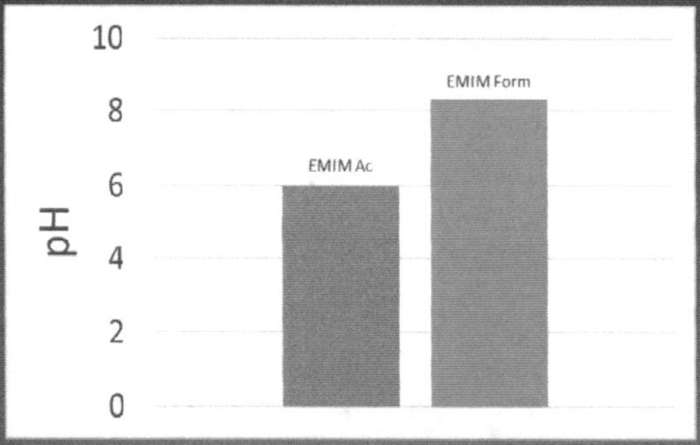

Higher pH solutions generally better at dissolving lignin

Dissolution of Lignin in Recalcitrant Biomass Using Ionic Liquid - Glycerol Mixtures AIChE Annual Meeting 2015 36

Viscosity

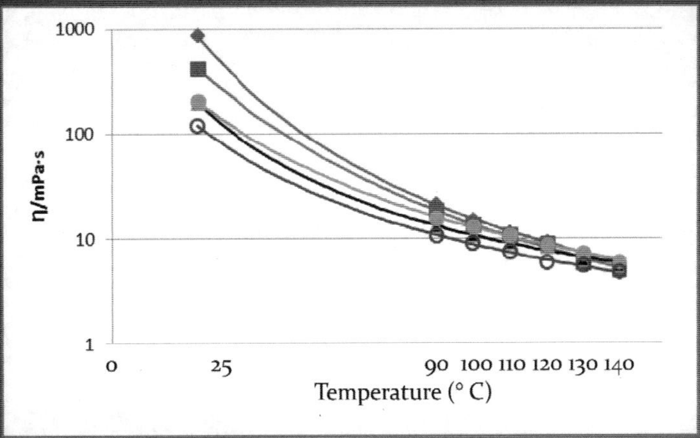

[C2mim][OAc] (EMIM Ac) mole fractions: ( ) 0 (all glycerol), ( ) 0.153 (25% IL), (▲) 0.351 (50% IL), ( ) 0.619 (75% IL), ( ) 1.0 (all EMIM Ac) .

Viscosity

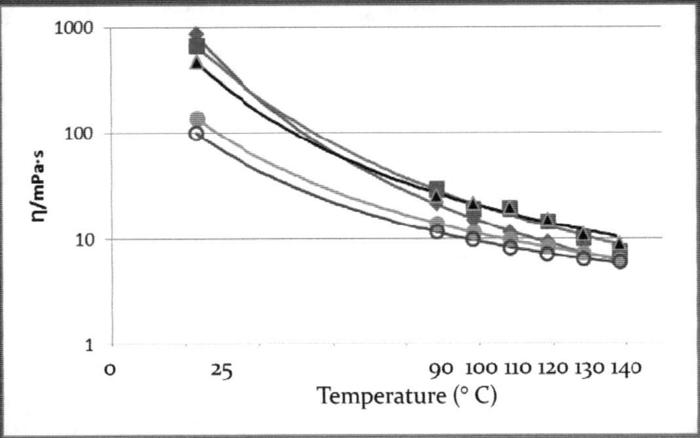

[C2mim][O2CH] (EMIM Form) mole fractions: ( ) 0 (all glycerol), ( ) 0.164 (25% IL), (▲) 0.371 (50% IL), ( ) 0.638 (75% IL), ( ) 1.0 (all EMIM Form) .

Viscosity

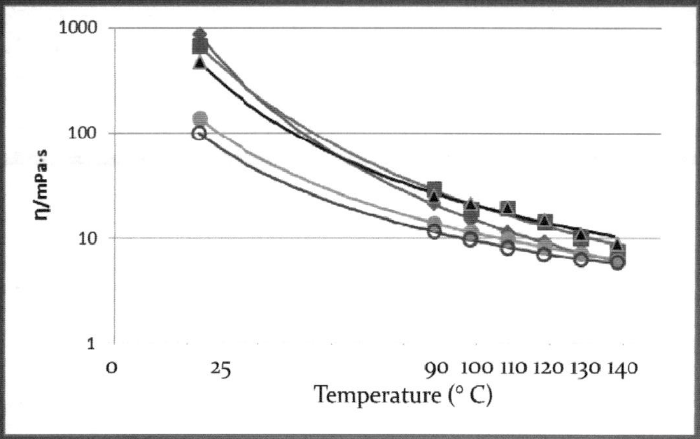

Is higher viscosity requiring greater energy for mixing causing particle shearing or fragmentation?

[C2mim][O2CH] (EMIM Form) mole fractions: ( ) 0 (all glycerol), ( ) 0.164 (25% IL), (▲) 0.371 (50% IL), ( ) 0.638 (75% IL), ( ) 1.0 (all EMIM Form) .

Density

- Excess Molar Volume, V_m^E – how volume of a mixture is different from what would be expected if there were no interaction between species.

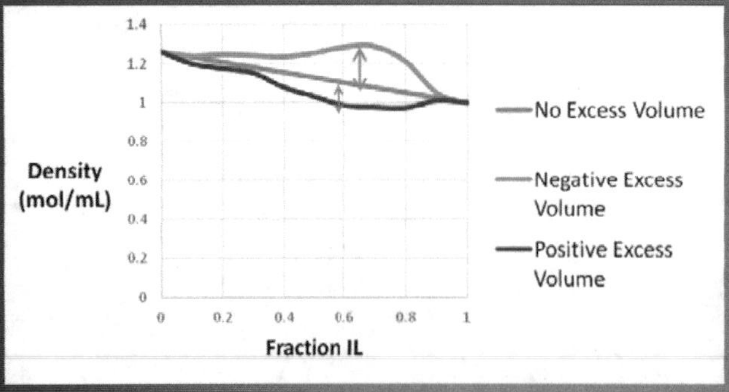

$$Vm^E = \frac{1}{molar\ density}$$

Less dense means greater volume per mole

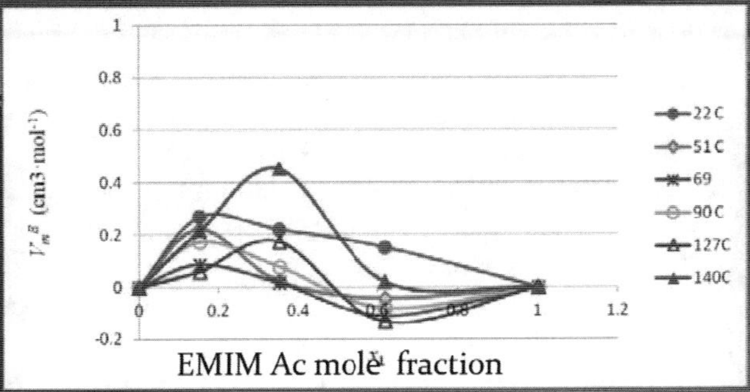

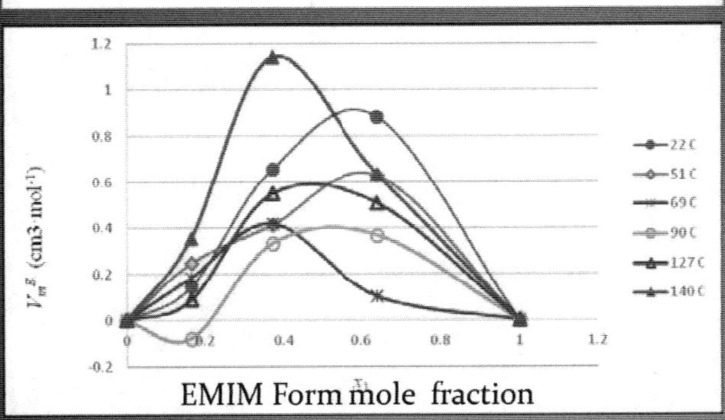

EMIM Ac – small V_m^E

EMIM Form – large, positive V_m^E
Particularly at 140° C

Species interactions effects on V_m^E *for IL with molecular solvent*

- ❖ Dissociation of positive and negative ions → V_m^E ↑
- ❖ Disruption of hydrogen bonding network in molecular solvent → V_m^E ↑

- ❖ Anions may hydrogen bond with the molecular solvent → V_m^E ↓
- ❖ Molecules from molecular solvent may fit in the interstices of IL → V_m^E ↓

EMIM Ac has smaller V_m^E and less positive V_m^E :
↓ *effects &* ↑ *effects may counteract each other*

EMIM Form has larger V_m^E and positive V_m^E ↑ *effects outweigh* ↓ *effects*

Why does EMIM Ac not work as well as EMIM Form with glycerol added?

Glycerol may hydrogen bond with Ac anions, so they cannot penetrate biomass structure.

Tighter structure

Positive V_m^E means species have less attractive interactions and a loose structure.

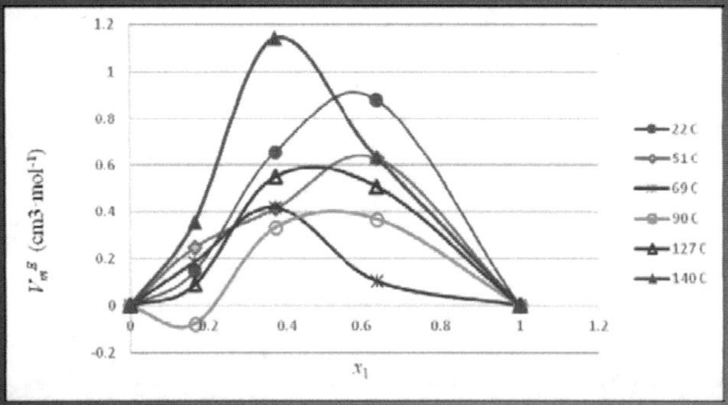

EMIM Form

Why does EMIM Form work with glycerol added?

Positive V_m^E allows anion to enter holes in the biomass.

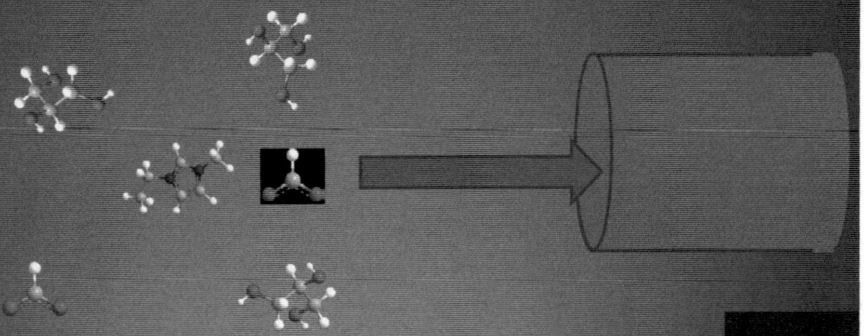

Loose structure

Conclusions

- ❖ IL – Glycerol mixtures can be used to pretreat recalcitrant biomass for hydrolysis to glucose and hemicellulosic sugars
- ❖ Higher temperatures and more equal IL: Glycerol ratios may increase excess molar volume in the mixtures
- ❖ Higher molar volumes may increase lignin removal, enhancing sugar yields

Future Work

- ❖ Optimize temperature, dissolutions times, and particle diameters
- ❖ Pretreat corn stover or other easier-to-deconstruct biomass with IL - glycerol mixtures

Acknowledgments

Thank you!

Dr. Charles J. Coronella, adviser.

Dr. Victor Vasquez, Dr. Hongfei Lin, Dr. Claus Tittiger, and Dr. John Anderson.

Co-workers: Dr. Toufiq Reza, M. Helal Uddin, Dr. Lisha Yang, Xiaokun Yang, Mi Lu, Genica Chow, and Philip Hyland.

Support and Inspiration from: Dr. Alan Fuchs, Dr. Ravi Subramanian, Dr. Kent Hoekman, Dr. John Cushman.

Graduate Student Association of UNR and USDA NIFA for financial support.

USDA United States Department of Agriculture National Institute of Food and Agriculture

This material is based upon work supported by the National Institute of Food and Agriculture, U.S. Department of Agriculture, under Agreement No. 2013-67011-21011

Any opinions, findings, conclusions, or recommendations expressed in this publication are those of the author(s) and do not necessarily reflect the view of the U.S. Department of Agriculture.

Acknowledgments

Thank you!

Dr. Charles J. Coronella, adviser.

Dr. Victor Vasquez, Dr. Hongfei Lin, Dr. Claus Tittiger, and Dr. John Anderson.

Co-workers: Dr. Toufiq Reza, M. Helal Uddin, Dr. Lisha Yang, Xiaokun Yang, Mi Lu, Genica Chow, and Philip Hyland.

Support and Inspiration from: Dr. Alan Fuchs, Dr. Ravi Subramanian, Dr. Kent Hoekman, Dr. John Cushman.

Graduate Student Association of UNR and USDA NIFA for financial support.

USDA United States Department of Agriculture National Institute of Food and Agriculture

This material is based upon work supported by the National Institute of Food and Agriculture, U.S. Department of Agriculture, under Agreement No. 2013-67011-21011

Questions?

Dissolution of Lignin in Recalcitrant Biomass Using Ionic Liquid - Glycerol Mixtures AIChE Annual Meeting 2015

49

422289 Characterization of Lignins Produced from Woody Biomass Using Solvent Extraction

Abstract:

Abstract Submission to Be Considered as an Oral Presentation at ***AICHE 2015Annual Meeting*** in *T4A01 Biomass Characterization, Pretreatment and Fractionation I*

Characterization of Lignins Produced from Woody Biomass Using Solvent Extraction

Ruoshui Ma[1], Carlos Alvarez-Vasco[1], Mond Guo[1], Vincent Hebert[1], Jinwen Zhang[1], Xiao Zhang[1]

1. Washington State University, Richland, WA, 99354

Lignin is the largest nature source of renewable material with an aromatic skeleton and thus a potential source for a range of value-added chemicals/products. Developing novel solvent systems for selectively extracting lignin from biomass and for lignin conversion to value-added chemicals have gained increasing interest recently.

In this presentation, we will report current progress from our group on developing a new type of solvent for selective remove of lignin from woody biomass including Douglas fir and Poplar. Detailed characterization of lignin properties and investigation of its potential for conversion to value-added chemicals will be presented.

Presenting author: Ruoshui Ma, Chemical Engineering, Washington State University, Richland, WA, 99354

Contact: ruoshui.ma@email.wsu.edu

429326 Changes in Lignin Structure after Autohydrolysis Pretreatment Analyzed By Quantitative NMR Techniques

Abstract:

The economic success of a lignocellulosic biorefinery is highly sensitive to the final price at which the lignin residue can be sold for at market. To identify possible pathways for lignin valorization, it is first necessary to understand the changes to the chemical structure of lignin during the conversion process. It is known that autohydrolysis pretreatment will alter lignin structure through both depolymerization and repolymerization reactions. These reactions were tracked by comparing the chemical structures of isolated lignin fractions from both raw and autohydrolyzed biomasses using quantitative 13C and 2D-HSQC NMR. Three hardwood and three non-wood lignocellulosic biomasses were subjected to identical autohydrolysis pretreatments at 180 °C for 40 minutes. A lignin isolation scheme involving the generation of milled-wood lignin (MWL) and cellulolytic enzyme lignin (CEL) was applied to both the raw and autohydrolyzed biomasses, which were then analyzed by NMR. For hardwood *E. nitens,* it was found that the amount of β-O-4 ether bonds decreased from 53 per 100 aromatic units in the raw MWL to 34 per 100 aromatic units in the autohydrolyzed CEL, showing evidence of acidolysis depolymerization reaction occurring in the lignin. Additionally, the amount of cinnamyl alcohol end groups increased after autohydrolysis from 1 to 4 per 100 aromatic units, suggesting additional depolymerization reactions involving liberation of these moieties. The quantification of major inter-lignin linkages in the remaining hardwood and non-wood lignins will be presented. In addition, we will discuss the challenges and our approach of non-wood lignin characterization induced by the *para*-coumaroyl and feruloyl linkages within lignin as well as between lignin and hemicellulose.

430286 Extractives of Douglas-fir and their Implications for Fermentation Based Biofuel Production

Monday, November 9, 2015: 10:35 AM
250C (Salt Palace Convention Center)

Karl Oleson, Kayla Sprenger, Jim Pfaendtner and Daniel T. Schwartz, Chemical Engineering, University of Washington, Seattle, WA

Douglas fir (*Pseudotsuga menziesii*) has been investigated as a feedstock for biofuel processes due to its abundance in western North American timberland. This plentiful feedstock has the potential to be a large component of a forestry residue based biofuel industry. Residues from timber harvest and forest thinnings (sometimes called "slash") are comprised of branches, broken or defective tree parts, tops, and trees not meeting grade specifications. This material is often considered unmerchantable and so it is collected, piled, and burned (or left to decay) after forestry operations to reduce fire hazards and improve forest stands. Routing Douglas fir residues into biofuel production offers the potential for a sustainable low carbon fuel.

To design economical biofuel facilities using forestry residues, a better understanding of how chemicals components in the feedstock affect biofuel process steps is needed. Diverse extractive compounds make up 5% to 25% of the dry weight for different tissues of Douglas fir [1,2], but are rarely accounted for in biofuel studies. Despite the well-documented role of specific extractive components in the pulp and paper literature, little detailed attention has been focused on extractive molecules in biofuel processing. Instead, it is typical to define a lumped species with a single property set, a psuedo-extractive, that is tracked in process simulations. Our studies provide a starting point for understanding the effects of molecularly distinct extractives in fermentation-based biofuel production.

We create a molecularly representative model of Douglas fir logging slash by drawing from a wide range of literature. Our feedstock model extractive components consist of 3.3% (w/w) polymeric polyphenols, 2.4% flavonoids, 1.6% fats and waxes, 0.4% diterpenes, 0.2% phytosterols, and 0.1% monoterpenes on a dry (no water) basis. See Figure 1 for representative molecules of these extractive classes [2-6]. ASPEN simulations are used to track how extractives move through a biofuel process based on sulfite/bisulfite pretreatment, saccharification, and fermentation. Steps such as pretreatment, pressing to remove spent sulfite liquor, enzymatic hydrolysis, and evaporation are simulated and molecules of interest are identified in key streams. Unlike process simulations that lump all extractives into a single psuedo-species, we find different classes of extractive molecules preferentially partitioning between the vapor, product, and waste streams.

Our simulation results indicate that the saccharification step may be particularly susceptible to inhibition from extractives. Saccharification is inhibited by noncellulose components binding to cellulose as well as binding to the hydrolyzing enzymes. Astringent molecules known to bind to cellulases such as the condensed tannins are very abundant in feedstocks containing bark and our simulation indicates they appear in significant quantities in the feed to saccharification. As enzymes are still one of the most expensive steps in biofuel processing, ways to use lower titers of enzymes or bioengineer more effective enzymes need to be investigated. Knowledge of the extractive inhibitors can help guide this investigation and subsequent optimization.

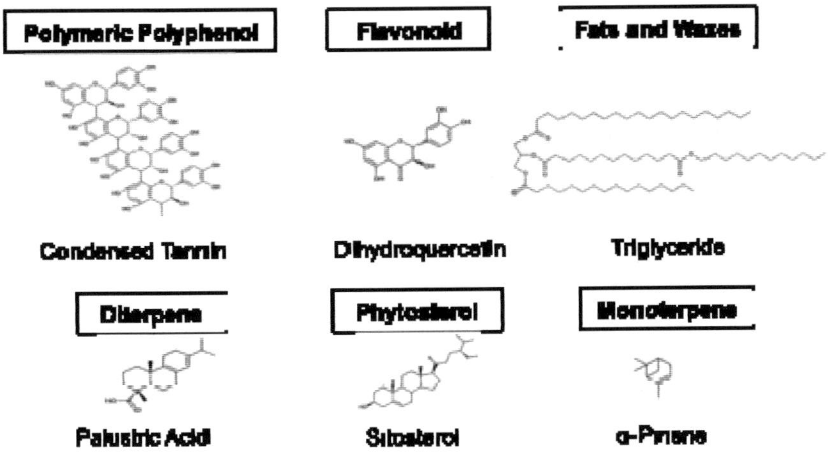

Figure 1: The six most abundant classes of extractives in Douglas fir and molecules representative of their class.

Molecular dynamics simulations of cellulase-extractives systems are conducted using GROMACS. These simulations identify which extractives bind to different cellulase active sites, and the stability of extractive-enzyme complexes. The cellulases are modeled in aqueous media and include a cellobiohydrolase from *Hypocrea Jecorina*, cellobioase from *Aspergillus Niger*, and an endoglucanase from *Acidothermus Cellulolyticus*. Both the catalytic domain and carbohydrate-binding module are simulated for the cellobiohydrolase and endoglucanase. The extractives used include condensed tannin, dihydroquercetin, a fatty acid, palustric acid, sitosterol, and α-pinene. Figure 2 shows dihydroquercetin binding onto the carbohydrate-binding module of the cellobiohydrolase.

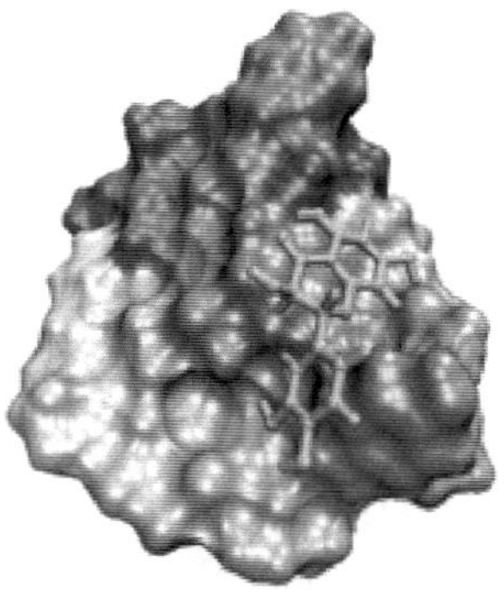

Figure 2: The extractive dihydroquercetin (green) is shown binding to the carbohydrate-binding module of the cellobiohydrolase from *Hypocrea jecorina* (colored by "index", based on amino acid location in the peptide sequence).

This work identifies the common extractives in Douglas fir, models how they propagate in important process streams, and assesses their affect on saccharification. These studies significantly improve our knowledge of the effect extractives have on biofuel processes as well as lay the groundwork for further studies of Douglas fir forestry residue.

References

[1] Kaar WE, Brink DL. Summative Analysis of 9 Common North American Woods. J Wood Chem Technol 1991;11(4):479-94.

[2] Kurth EF. Chemicals from Douglas-Fir Bark. TAPPI 1953;36(7):119A-22A.

[3] Dellus V, Mila I, Scalbert A, Menard C, Michon V, duPenhoat C. Douglas-fir Polyphenols and Heartwood Formation. Phytochem 1997;45(8):1573-8.

[4] Foster DO, Zinkel DF, Conner AH. Tall Oil Percursors of Douglas fir. TAPPI 1980;63(12):103-5.

[5] Erdtman H, Kimland B, Norin T, Daniels PJL. The Constituents of the "Pocket Resin" from Douglas Fir Pseudotsuga menziesii (Mirb.) Franco. Acta Chem Scand 1968;22(3):938-42.

[6] Fischer F, Koch H, Borchers B, Hontsch R, Pruzina KD. Preparation and Use of Phytosterols from Wood. Pharm 1981;36(7):456-62. ADDIN EN.REFLIST

418207 Supercritical Assisted Pretreatment of Cassava and Plantain Residues

Abstract:

Daza Laura[1], Betancourt Ramiro[1], Cardona Carlos A[1**].

[1] Universidad Nacional de Colombia sede Manizales, Instituto de Biotecnología y Agroindustria, (+57) (6) 879400 ext 55354.

[**] Corresponding author: ccardonaal@unal.edu.co

Colombia is a tropical country assorted of soils and hydric resources which produces a wide variety of crops. The agricultural activities generates high amounts of waste biomass. Cassava stem and Plantain pseudostem are lignocellulosic residues generated during harvesting stage. The interest and use of these type of residues as raw materials for biorefineries to produce chemicals, biofuels and other products has been increasing as reported by authors as *Moncada* et al [1], [2]. The available amounts and chemical composition are some advantages of this group of raw materials. Nevertheless, the morphologic arrangement of cellulose, hemicellulose and lignin becomes in the main drawback for enzymatic and microorganism attack. Pretreatments are used to reduce the cellulose crystallinity as well as the hemicellulose and lignin content.

The conventional pretreatments involves the use of hydrothermal, acid, alkaline and organosolv pretreatments [3]. Non-conventional pretreatment includes supercritical H_2O, ammonia and CO_2among others [4]. The pretreatments using supercritical carbon dioxide in presence of moderate to high moisture levels produce carbonic acid and generates biomass swelling. The produced carbonic acid cause a hydrolytic effect in the biomass [5].
Cassava stem and Plantain pseudostem were experimentally characterized by measuring moisture content (AOAC 928.09 method), Klason lignin content (TAPPI 222 om-83 method), acid-soluble lignin content (TAPPI 250UM-85 method) holocellulose content (ASTM Standard D1104 method), cellulose content (TAPPI 203 os-74 method), and ash content (TAPPI Standard T211 om-93 method). Then, the selected raw materials were submitted to a non-conventional and conventional pretreatment. Supercritical carbon dioxide was used as non-conventional pretreatment. The pressure, temperature and moisture content were experimentally evaluated. The conventional pretreatment was developed using 2% sulfuric acid. Later on, the pretreated samples were submitted to enzymatic hydrolysis to produce glucose by cellulases enzymes. The sugars, and inhibitory compounds yields were determined through spectrophotometric analysis.

The sugars production from lignocellulosic biomass using supercritical carbon dioxide and sulfuric diluted acid (case base) pretreatments were evaluated using the commercial software Aspen Plus V8.2 (ASPEN TECHNOLOGY INC). The economic evaluation was performed using the software commercial Aspen Process Economic Analyzer V8.2 (ASPEN TECHNOLOGY INC) taking into account the Colombian context with an annual interest rate of 16% and an income tax of 25%, with electricity and water costs according to the case study conditions. Furthermore, the evaluation was realized for a period of 10 years using as depreciation method the straight-line. Finally, the benefits provided by supercritical carbon dioxide pretreatment reduces the inhibitory compounds released compared to the conventional sulfuric diluted acid. However, the techno-economic analysis demonstrated that supercritical fluids pretreatment is still an expensive option in this case.

References

[1] J. Moncada, J. a. Tamayo, and C. a. Cardona, "Integrating first, second, and third generation biorefineries: Incorporating microalgae into the sugarcane biorefinery," *Chem. Eng. Sci.*, vol. 118, pp. 126–140, 2014.

[2] L. E. Rincon, J. Moncada Botero, and C. A. Cardona Alzate, *Catalytic Systems for Integral Transformations of Oil Plants through Biorefinery concept.*, First Edit. Manizales, Colombia: Universidad Nacional de Colombia sede Manizales, 2013.

[3] C. E. Wyman, S. R. Decker, M. E. Himmel, J. W. Brady, and C. E. Skopec, "Hydrolysis of Cellulose and Hemicellulose," in *Polysaccharides Structural Diversity and Functional Versatility*, Second., S. Dimitriu, Ed. New York, 2005.

[4] M. Gao, F. Xu, S. Li, X. Ji, S. Chen, and D. Zhang, "Effect of SC-CO2 pretreatment in increasing rice straw biomass conversion," *Biosyst. Eng.*, vol. 106, no. 4, pp. 470–475, 2010.

[5] N. Narayanaswamy, A. Faik, D. J. Goetz, and T. Gu, "Supercritical carbon dioxide pretreatment of corn stover and switchgrass for lignocellulosic ethanol production," *Bioresour. Technol.*, vol. 102, no. 13, pp. 6995–7000, 2011.

421613 New Trends to Improve the Quality of Recycled Papers

Abstract:

The consumption of recovered paper has been in continuous growth during the past decades. This development has been boosted by technological progress due to the environmental awareness, saving energy, population growth and the shortage of wood supply. Furthermore, the demand of paper has decreased due to both the economic recession and the replacement of paper by other information supports and the good price competitiveness of recycled fiber decreases the production cost of the papermaking process.

Recovery rates have been increasing in North America for over a decade. In 2013 more than 63% of the paper and paperboard consumed in the U.S. was recovered. Nowadays, old corrugated containers (OCC) are the most significant category of recycled papers, based on the use (19.8 million tons in 2013) and recovery rate (88.5%) of this raw material.

However, by increasing the number of cycles in the fiber recycling, the mechanical properties of paper and board is gradually reduced which is mainly due to the decrease in the fiber bonding strength through the hornification phenomenon. In addition, the substitution of cellulose fibers by fillers in the paper stock due to their lower cost and their ability to improve optical properties in the final product also affect the mechanical properties of the paper. Two problems are usually associated with the use of the fillers. First, filler particles added to fibers suspended in water are not easily retained in the forming sheet, because there are often too small to be entrapped mechanically and because filler particles and fibers are negatively charged, so they repel each other. Second, filler particles interfere with the fiber-fiber bonding; therefore the linting problems in the offset printing can increase and furthermore other mechanical properties as tensile strength of the filled paper decrease. Synthetic retention systems such as cationic polyacrylamide (CPAM) and dual systems based in CPAM and microparticles are generally used to increase the interaction between fillers and fibers, leading to improved retention. However, the use of those retention systems can further contribute to decrease the mechanical properties of the product. Consequently, the overcoming of the second problem associated to the use of fillers is not easy.

On the other hand, customers are demanding products with better mechanical properties and quality. Consequently, during the last decades, many researches are focused on the feasibility of using different natural and/or synthetic polymers to improve the strength of paper and board made from recycled fibers; however, poor tensile strength and linting are still the main source of customer complaints to paper manufacturers. Therefore, new ways to improve the strength of recycled paper products should be explored to meet the industry's demand.

In this context, the use of nanofibrillated cellulose (CNF) has been studied to improve filler-fiber interaction and interfiber bonding and thus increasing tensile strength and reducing linting and dusting during printing process, since the mechanism of linting is generally a combined effect on the ink film splitting forces and the inter fiber bonding energy of the paper surface; at the same time that drainage and retention are improved.

The use of CNF presents some very interesting properties for papermaking such as a large specific surface area and high aspect ratio; moreover, it enhances tensile strength and reduces the porosity of the final paper sheet. Therefore, CNF is potential candidate for high filler-loaded papers and board as they are able to compensate for strength loss caused by the filler itself. However, the interaction of CNF and the filler during sheet forming is not yet well understood.

In this work, CNF was obtained from two different sources of cellulose: eucalyptus bleached pulp (E), and pine bleached pulp (P). Nanofibrillated material was obtained by TEMPO mediated oxidation, using 0.1 g NaBr, 0.016 g TEMPO and 5 mmol NaClO per each gram of pulp. The oxidation was performed at pH 10 by continuous addition of NaOH 0.5 M. Once the pulp was oxidized, a cleaning step was performed through filtration cycles using water. Finally, five steps of homogenization at 600 bar were applied to the oxidized pulp (1% consistency) in a laboratory homogenizer PANDA PLUS 2000 de GEA Niro Soavy (Parma, Italy). The yield in nanofiber material was higher than 95% in both CNF from the eucalyptus pulp (CNF-E) and CNF from pine pulp (CNF-P) and in both cases the transmittance at 800 nm was higher than 90%. The cationic demand and the polymerization degree of the CNF-E were the double than those for CNF-P.

Different dosages of additives and CNF (CNF-E and CNF-P) were individually added to the recycled furnish and the handsheet properties in terms of mechanical properties were investigated. In addition, new combinations were examined to find the optimal procedure for improving paper and paperboard properties. Moreover, the interaction between fillers and CNF was studied in order to determine their effects on drainage time, retention and strength properties of the handsheets. The results showed CNFs to have a potential to enhance the mechanical properties of the recycled paper and board they could be combined with traditional strength aids and with fillers for improving the quality of the final paper. Furthermore, the retention and drainage process were not significantly affected providing the dose of the retention aids is adapted to the presence of CNF and fillers.

In conclusion, the use of nanocellulose can open the door to the increase in the filler content of the paper and to the production of high quality recycled paper products, reducing the claims from customers to the manufacturers related to sheet failure and/or linting, both caused by poor interfiber bonding.

The authors wish to thank the Economy and Competitiveness Ministry of Spain for the support of the projects with references CTQ2012-36868-C02-01 and CTQ2013-48090-C2-1-R.

The Effects of Additives on the Optical and Mechanical Properties of Cellulose Nanocrystal Films

AUBURN UNIVERSITY
SAMUEL GINN COLLEGE OF ENGINEERING

Joshua Passantino
Alexander Haywood
Virginia A. Davis
Department of Chemical Engineering

November 9, 2015

Outline

- Objectives & motivation
- Issues with cellulose nanocrystal films
- Theory
- Film creation process
- Results
- Conclusions
- Future work

2

AUBURN UNIVERSITY
SAMUEL GINN COLLEGE OF ENGINEERING

Objectives

- Compare fabrication methods for cellulose nanocrystal (CNC) films
  - Polymer additives
  - Shear
  - Flocculation
- Prove mechanical anisotropic properties of sheared CNC films

3

AUBURN UNIVERSITY
SAMUEL GINN COLLEGE OF ENGINEERING

Motivation

- Cellulose nanocrystal films:
  - Films with desirable properties
  - Determination of best processing methods
- Applications include:
  - Optical sensors
  - One-time-use medical devices
  - Composite manufacturing

4

AUBURN UNIVERSITY
SAMUEL GINN COLLEGE OF ENGINEERING

Cellulose Nanocrystals (CNC)

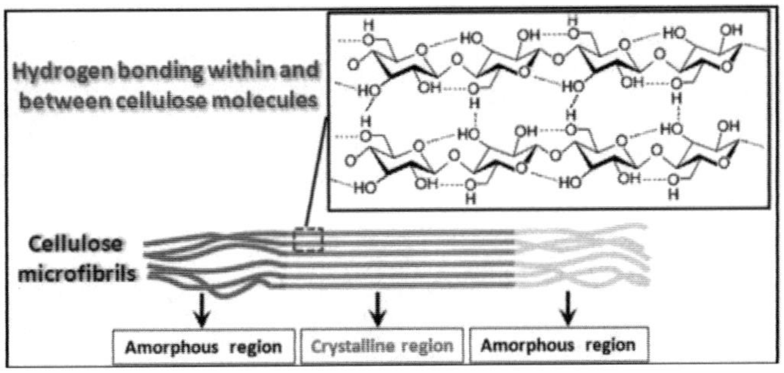

Sulfuric acid hydrolysis of cellulose (CNC-SA)

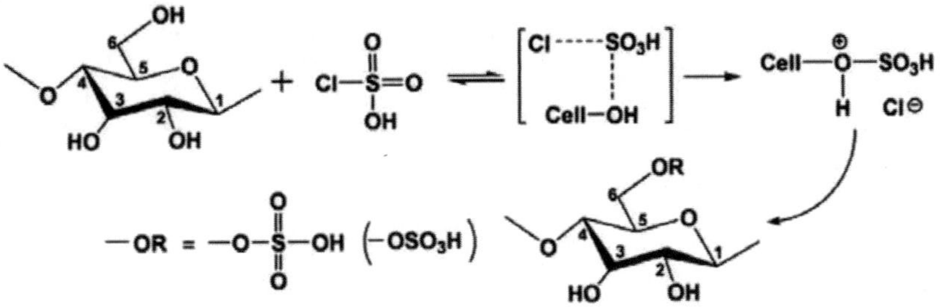

5 Image from: Chengjun Zhou and Qinglin Wu

AUBURN
UNIVERSITY
SAMUEL GINN
COLLEGE OF ENGINEERING

Shear in CNC Films

No shear **Shear**

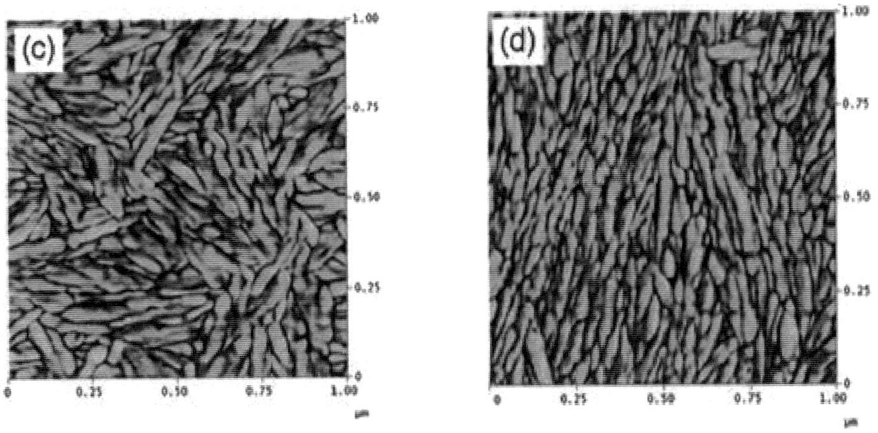

6 Image from: Wang, Haiyang, Yaozhuo Xu, Xinhong Yu, Rubo Xing, Jiangang Liu, and Yanchun Han. *Polymers*5.4 (2013): 1272-324. Web.

AUBURN
UNIVERSITY
SAMUEL GINN
COLLEGE OF ENGINEERING

Issues with pure CNC films

- Cracking
- Inconsistency
- Brittleness
- Thin films curl in on themselves
- Substrate interactions

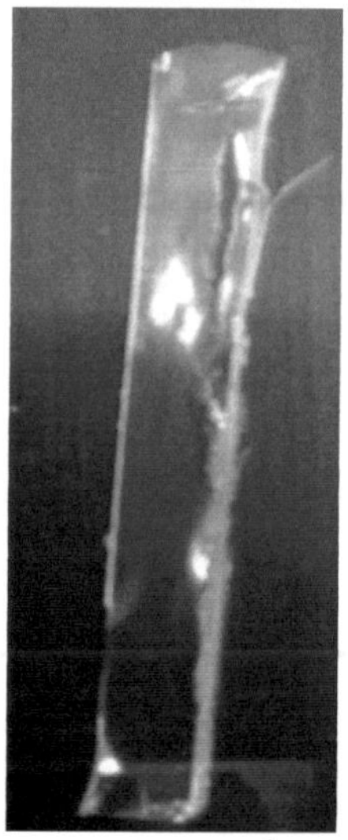

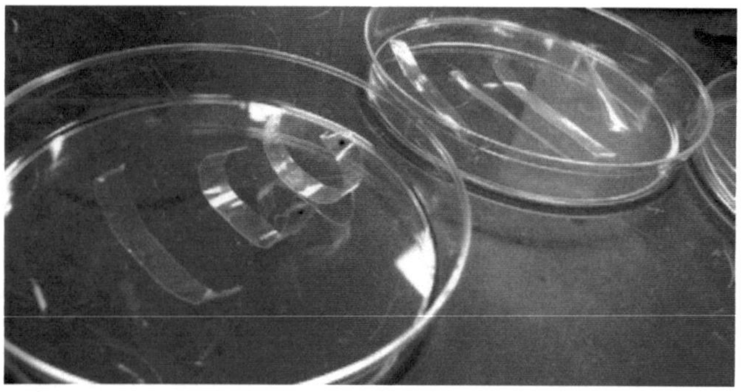

7

AUBURN UNIVERSITY
SAMUEL GINN COLLEGE OF ENGINEERING

Polymer Additives

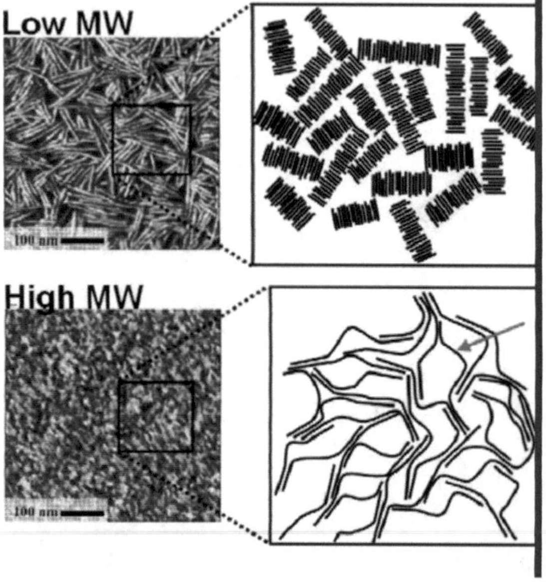

Polyethylene Oxide (PEO)

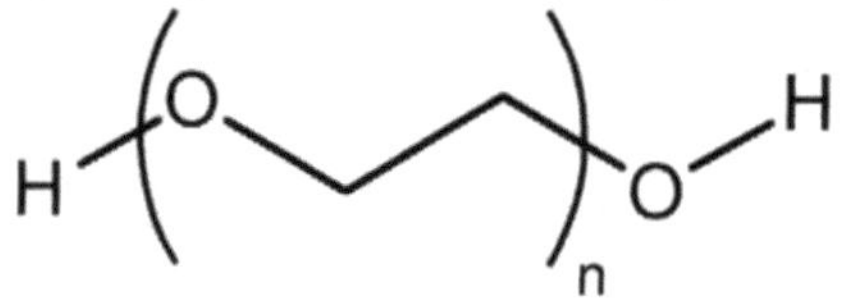

- Highly flexible
- Water soluble
- Used in common polymers
- Multiple molecular weights
  - 100,000
  - 1,000,000
  - 8,000,000

AUBURN UNIVERSITY
SAMUEL GINN COLLEGE OF ENGINEERING

8

Images from: Hoeger, Ingrid, Orlando J. Rojas, Kirill Efimenko, Orlin D. Velev, and Steve S. Kelley. *Soft Matter* 7.5 (2011): 1957. Web.

Flocculation

- Singh et al. (2009) reported that flocculation can help resolve cracking in films
- Flocs can be broken and reformed with mechanical shear
- Flocculation process
  - HCl releases H^+ ions that attach to negatively charged CNC-SA
  - CNC-SA is no longer negatively charged so it does not repel itself as strongly

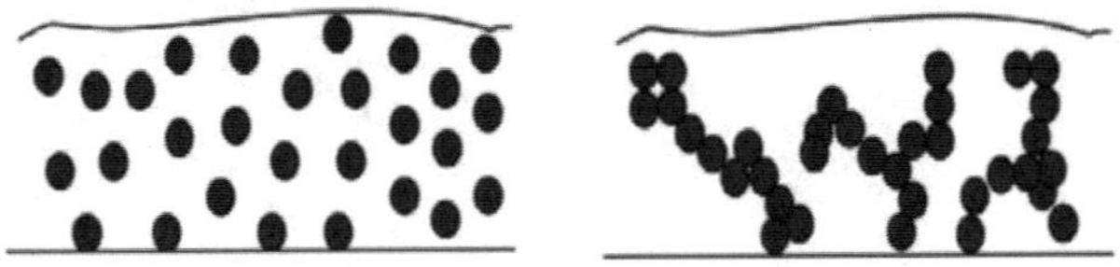

Not flocculated　　　　Flocculated

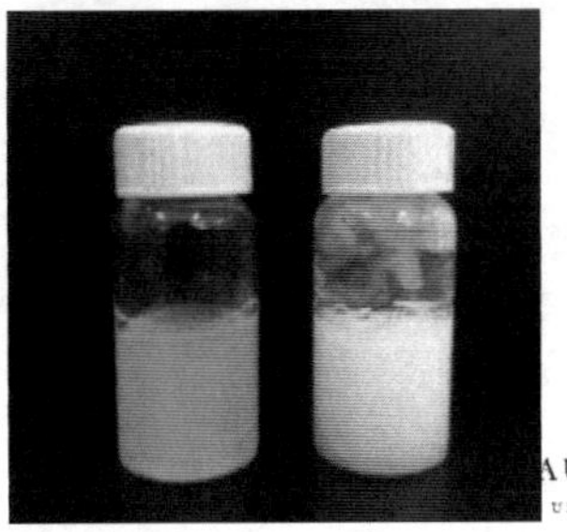

9　Image from: Singh et al. (2009)

Fabrication and Characterization

10

Film Creation

- Creation of CNC-SA solutions
  - Polymer mixed with 13 wt % CNC-SA then allowed to fully dissolve and disperse over night
  - Addition of HCl until the cellulose flocculates
  - Mixed with vortex mixer
  - Doctor blade coating

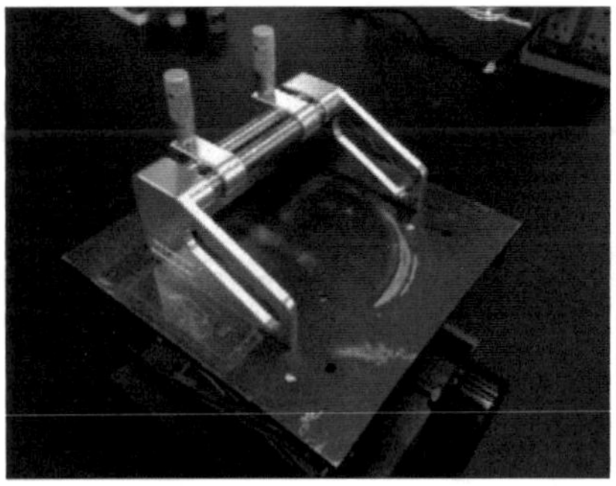

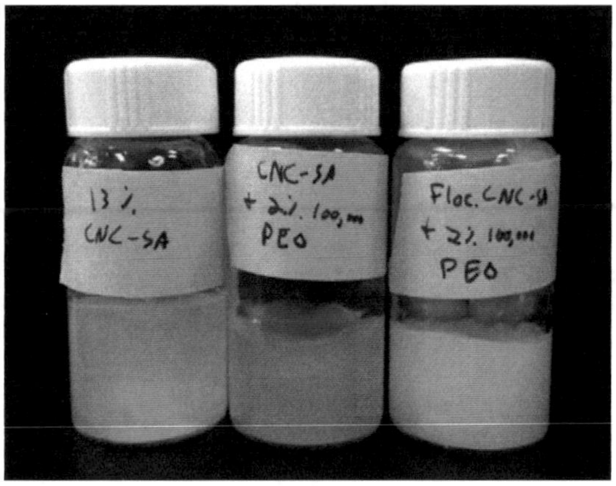

11

Optical Testing

- All images were taken at the center of the film
- Cross-polarized microscopy
  - Birefringence
  - Alignment

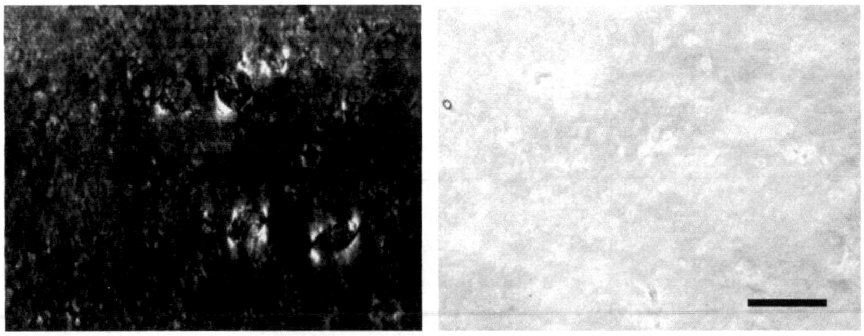

Well aligned film at 0° and 45°
Scale bar is 100 μm

12

AUBURN UNIVERSITY
Samuel Ginn College of Engineering

Tensile Testing

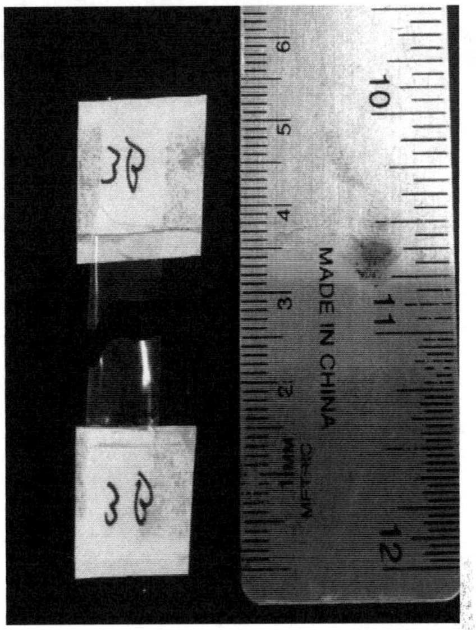

- INSTRON 5565
- Cross head speed: 0.5 mm/min
- Static load Cell: 100 N
- Paper holder
- Young's Modulus
- Tensile strength
- Toughness

AUBURN UNIVERSITY
Samuel Ginn College of Engineering

13

Results

AUBURN UNIVERSITY
Samuel Ginn College of Engineering

14

Initial Film Specifications

- Film thickness
  - A) Wet: 1 mm, Dry: 0.1 mm
  - B) Wet: 0.1 mm, Dry: 0.01 mm
- Shear rate
  - A) 2.6 s^{-1} for thick films
  - B) 26 s^{-1} for thin films
- Dry time: 5 hours in ambient conditions

AUBURN UNIVERSITY
Samuel Ginn College of Engineering

15

Thick films

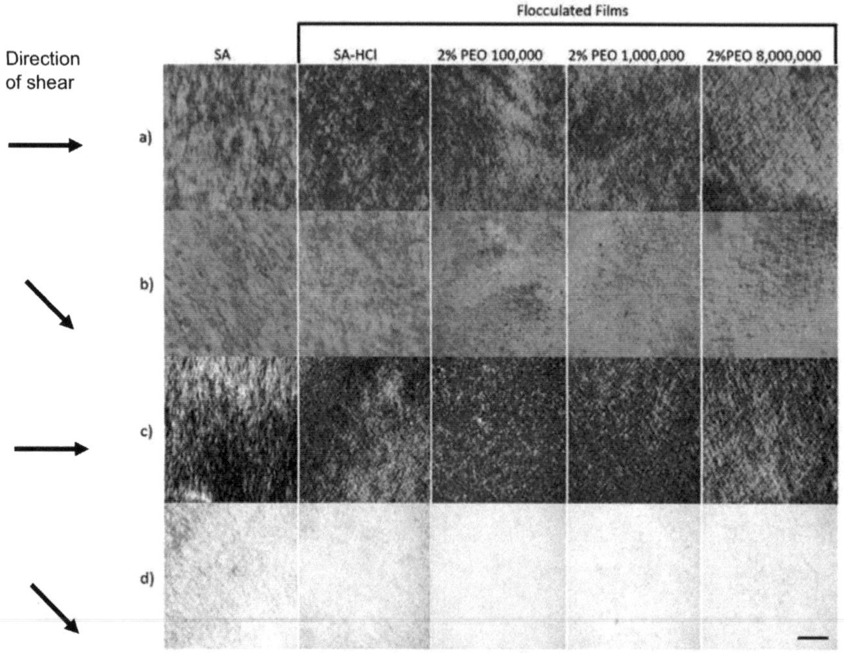

Scale bar is 200 μm

a & b: 0.1 mm thick films
c & d: 0.01 mm thick films

AUBURN UNIVERSITY
Samuel Ginn College of Engineering

16

Flocculated Films at Different Thicknesses

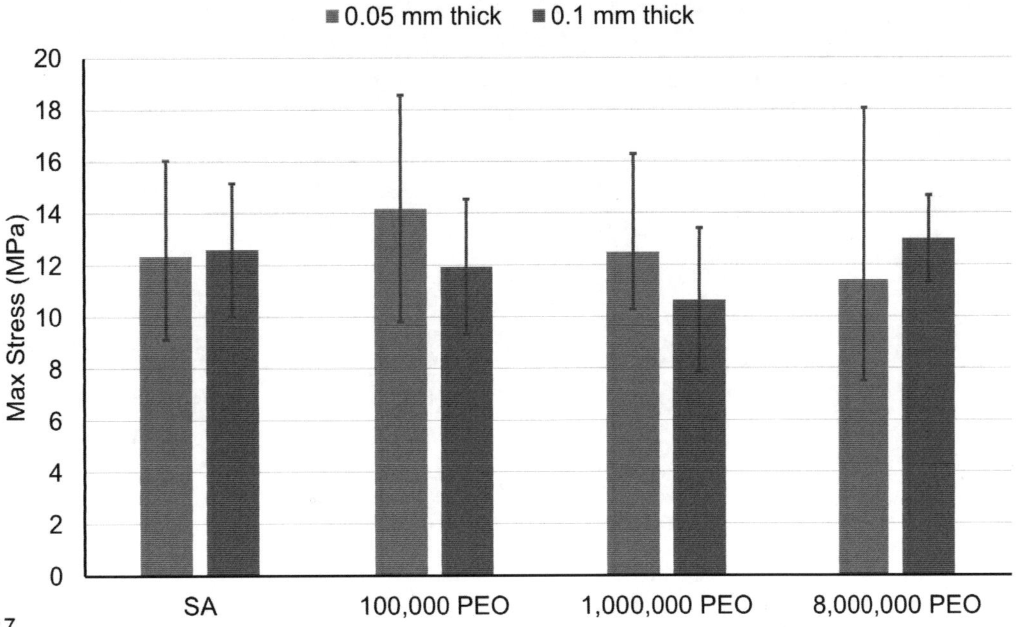

17

All error bars are the minimum and maximum for each film

New Film Specifications

- Film thickness
  - Wet: 0.5 mm, Dry: 0.05 mm
- Shear rate
  - 5 s^{-1} for all films
- Dry time: 5 hours in ambient conditions

18

AUBURN UNIVERSITY
SAMUEL GINN COLLEGE OF ENGINEERING

Parallel sheared films with additives

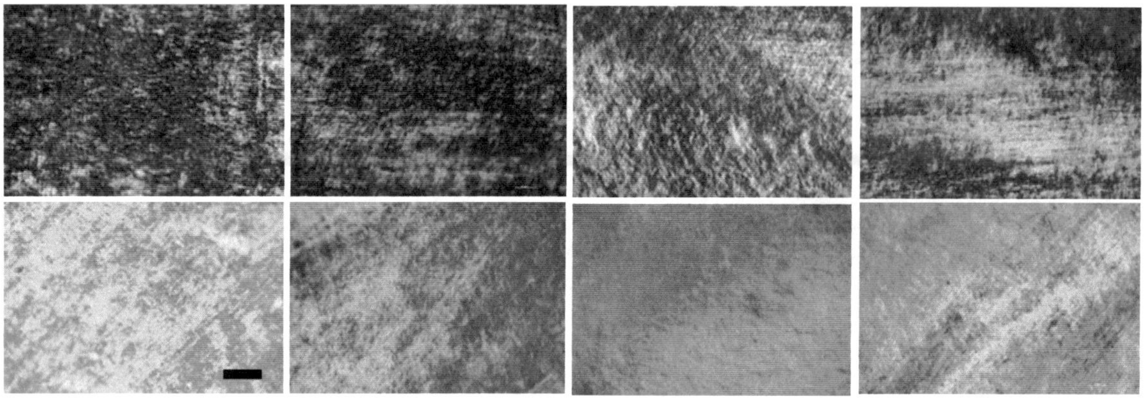

SA | 100,000 PEO | 1,000,000 PEO | 8,000,000 PEO

Scale bar is 200 μm

Top row shear direction

Bottom row shear direction

AUBURN UNIVERSITY
SAMUEL GINN COLLEGE OF ENGINEERING

19

Perpendicularly sheared films

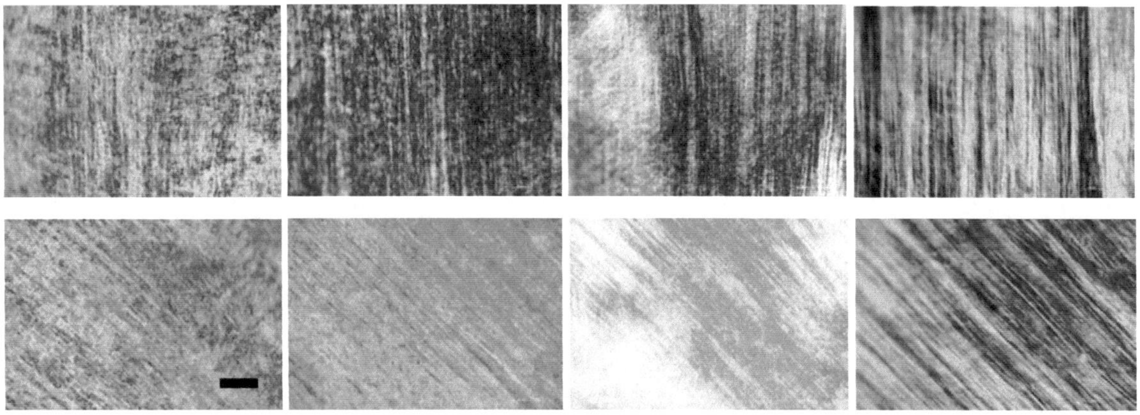

SA | 100,000 PEO | 1,000,000 PEO | 8,000,000 PEO

Scale bar is 200 μm

Top row shear direction

Bottom row shear direction

AUBURN UNIVERSITY
SAMUEL GINN COLLEGE OF ENGINEERING

20

Flocculated Films

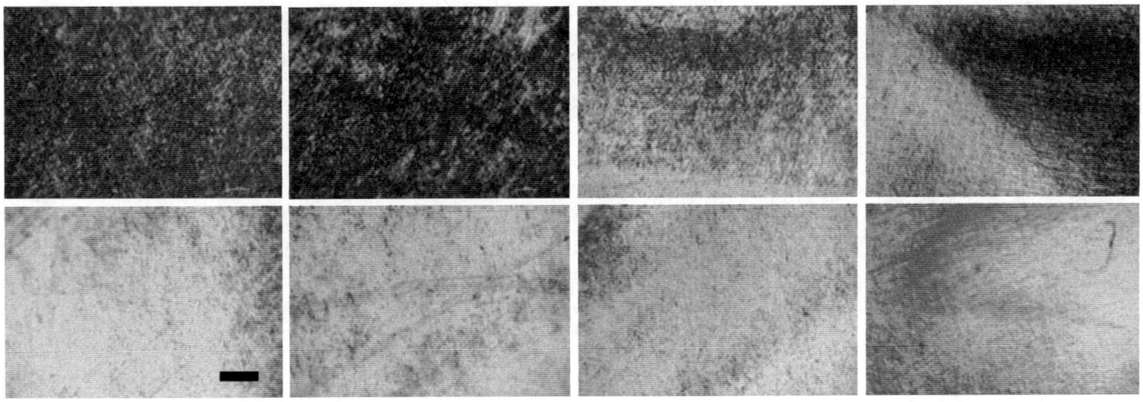

SA | 100,000 PEO | 1,000,000 PEO | 8,000,000 PEO

Scale bar is 200 μm

→ Top row shear direction

↗ Bottom row shear direction

AUBURN UNIVERSITY
SAMUEL GINN COLLEGE OF ENGINEERING

Comparison of pure CNC films

Scale bar is 200 μm

↓ Shear direction

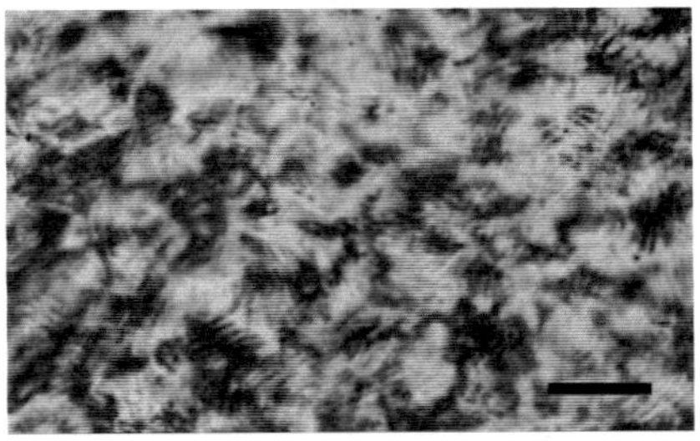

No shear, not flocculated

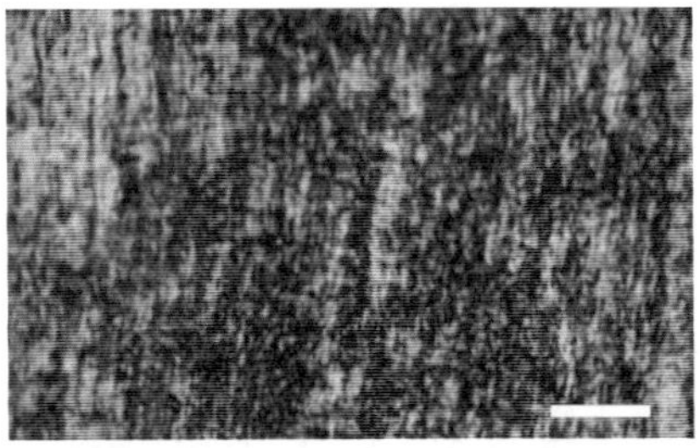

Parallel shear, not flocculated

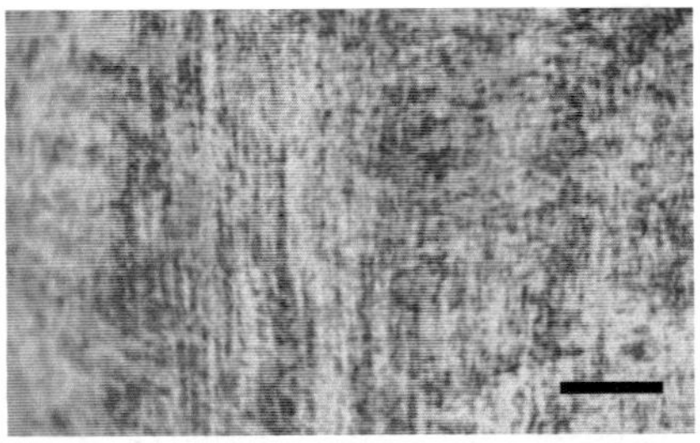

Perpendicular shear, not flocculated

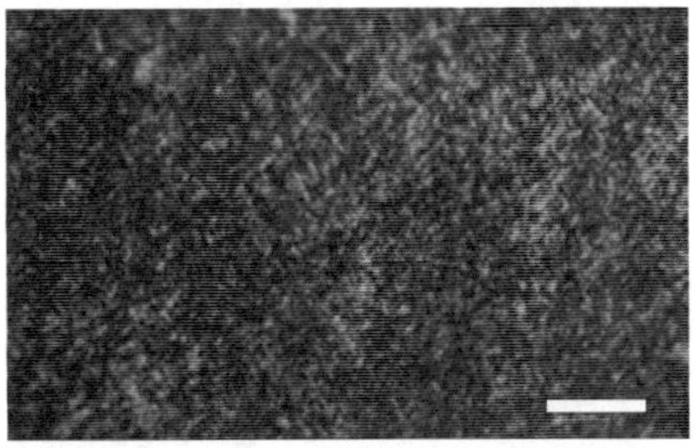

Parallel shear, flocculated

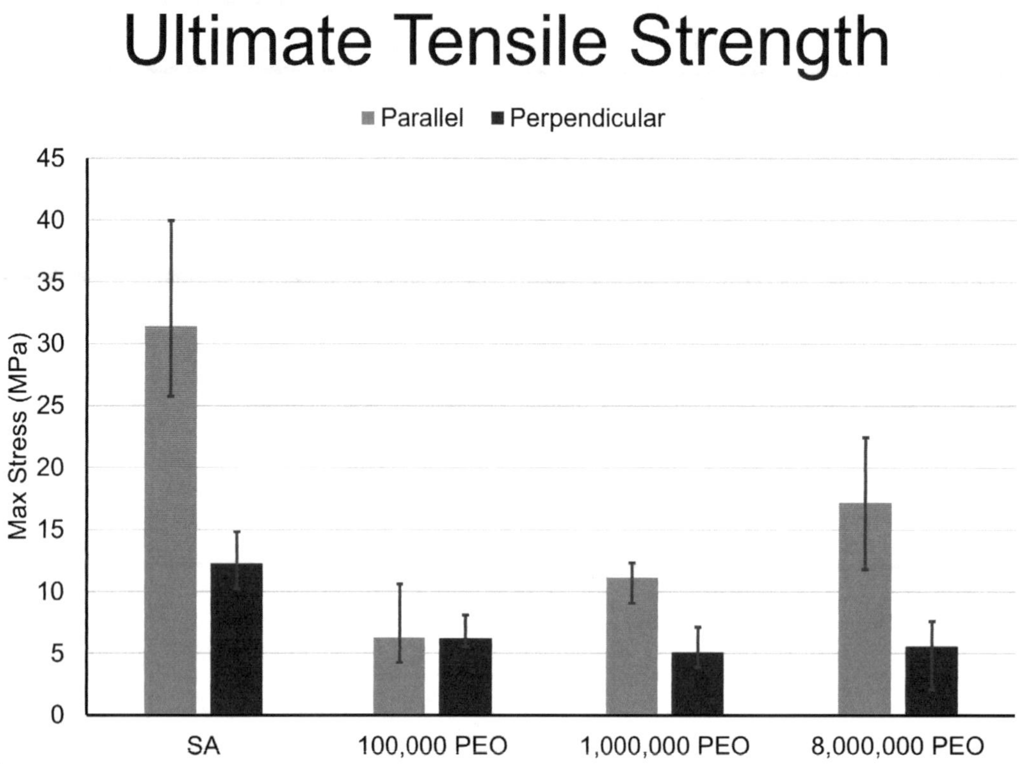

23 All error bars are the minimum and maximum for each film

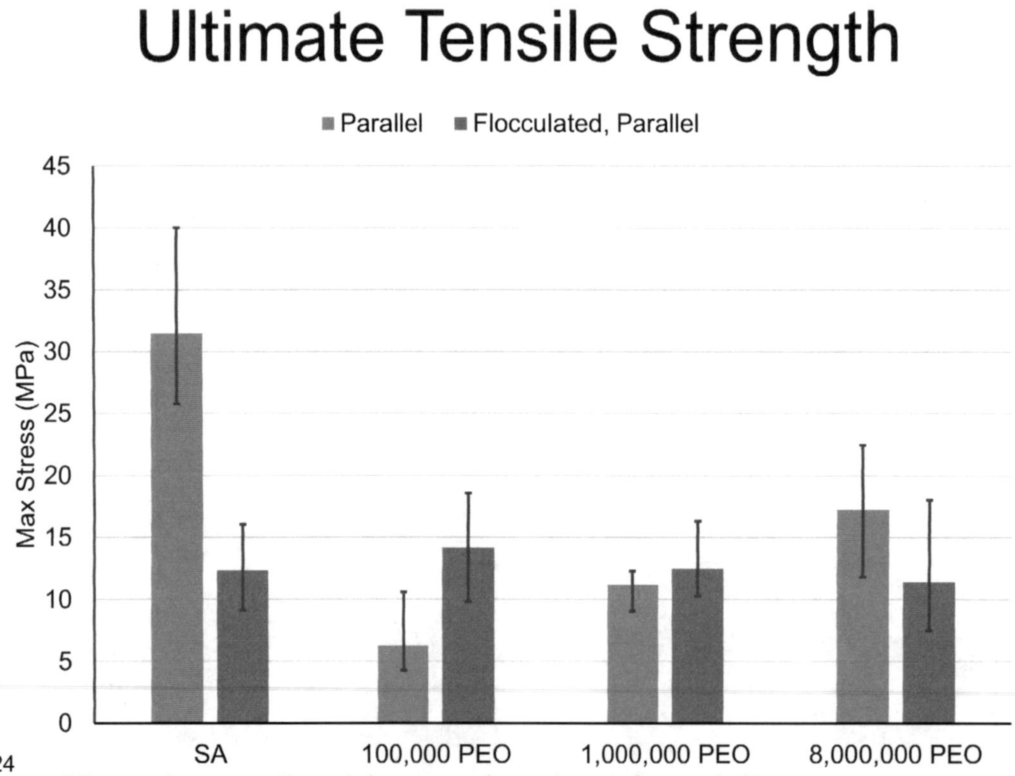

24 All error bars are the minimum and maximum for each film

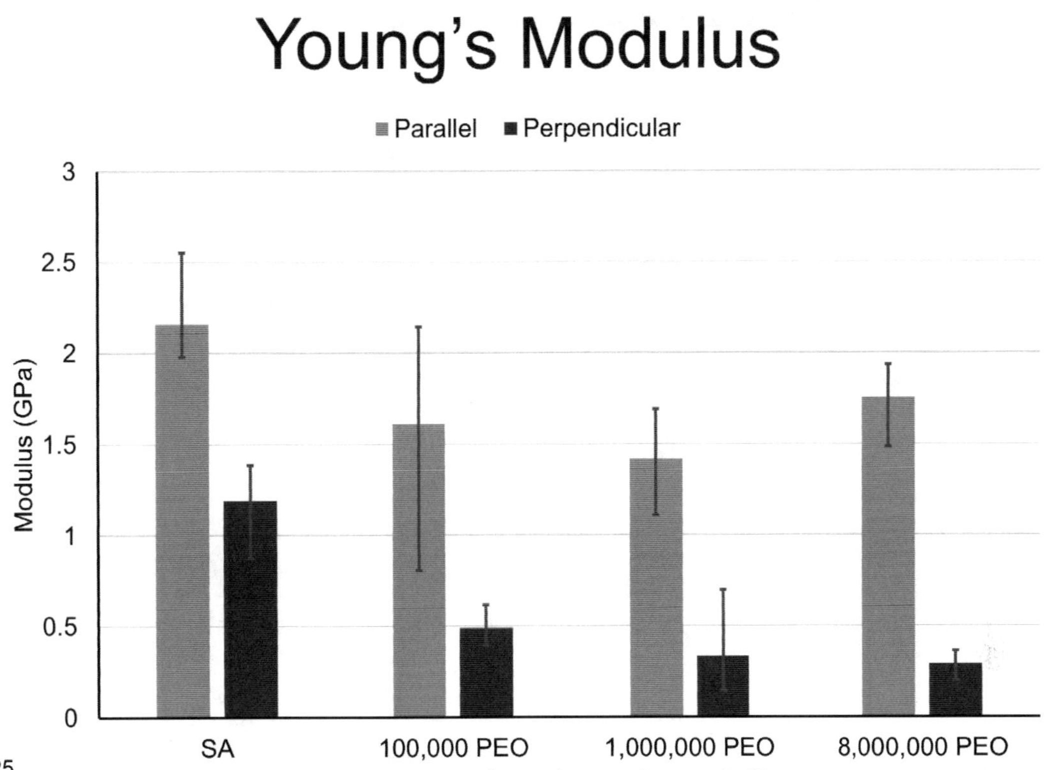

25 All error bars are the minimum and maximum for each film

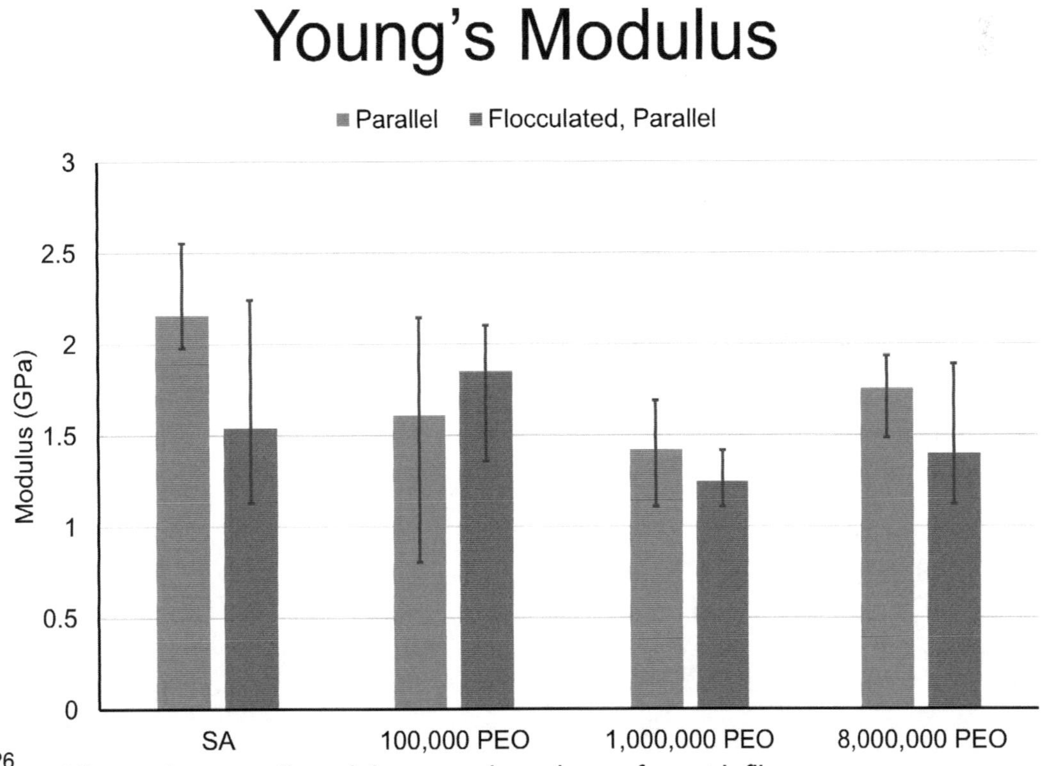

26 All error bars are the minimum and maximum for each film

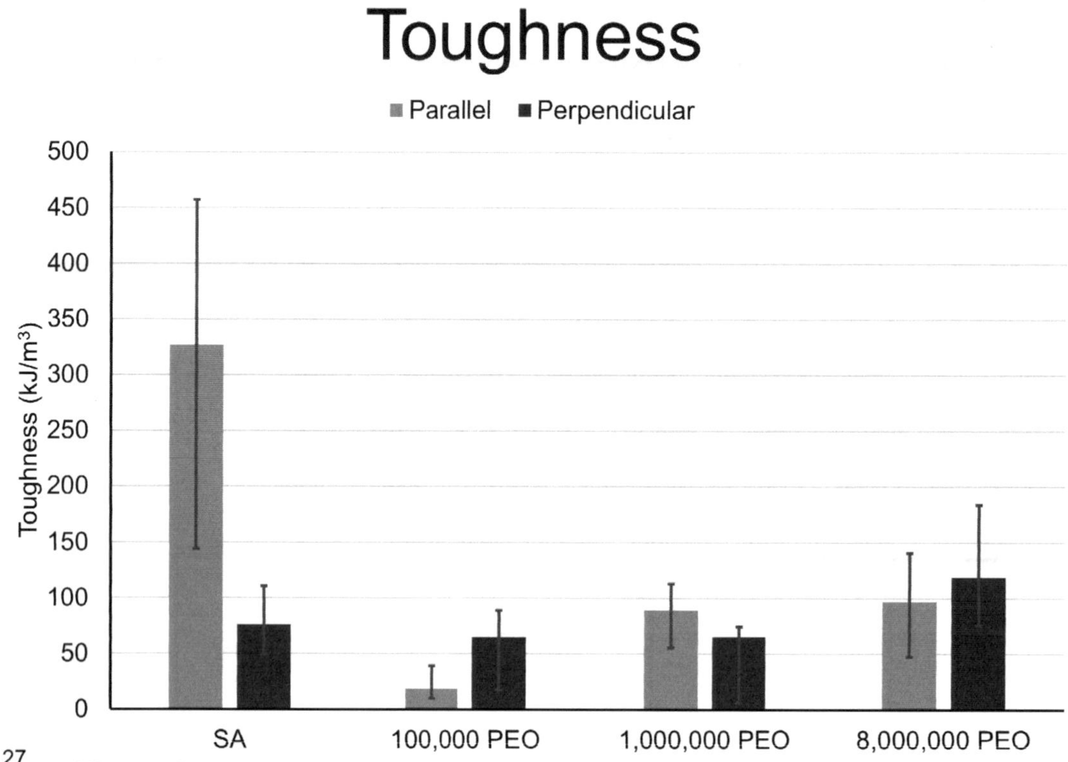

27 All error bars are the minimum and maximum for each film

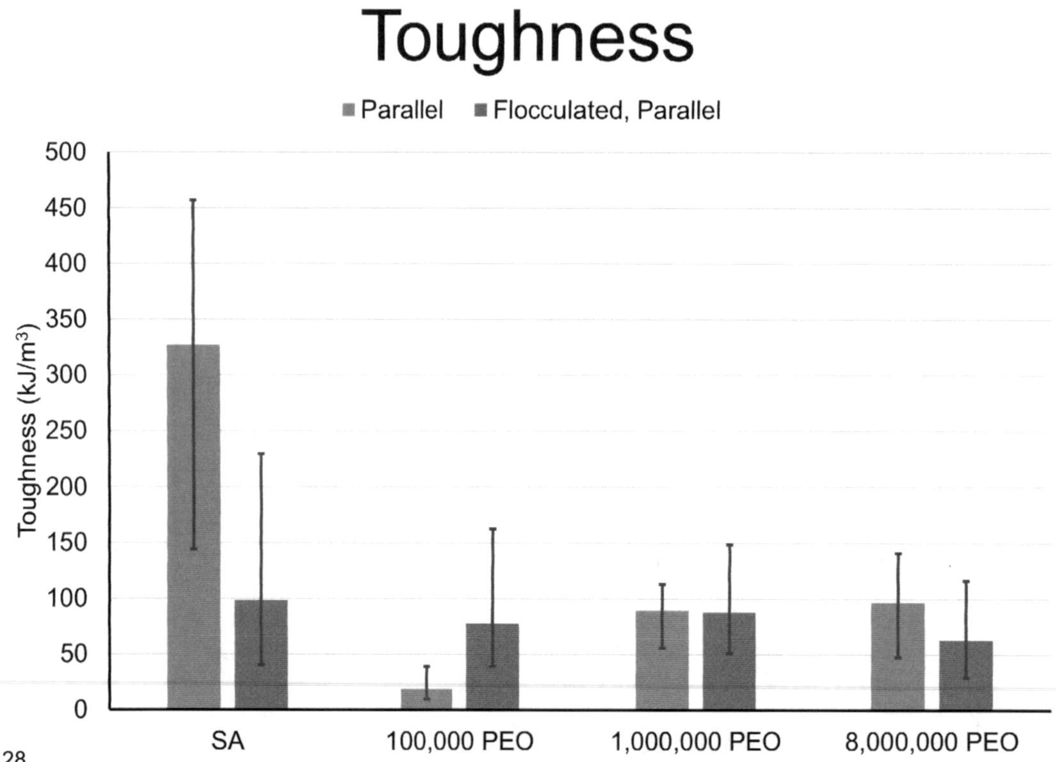

28 All error bars are the minimum and maximum for each film

Conclusions

- Addition of polymers reduces alignment and decreases strength, toughness, and the modulus
- Increases in MW of additives further reduce alignment ability, increase strength
- Pure CNC-SA is the strongest film of those tested
  - Highly variable
  - Thickness limited
- Parallel sheared films are stiffer and have higher tensile strengths than perpendicularly sheared films
  - Proof of mechanical anisotropy
- All parallel flocculated films have similar properties

AUBURN UNIVERSITY
SAMUEL GINN COLLEGE OF ENGINEERING

29

Future Work

- Effects of different additives
  - Different amounts of additives
- Comparison of nanoindentation and INSTRON data
- Altering shear rates
- Films from flocculated solutions with perpendicular shear

AUBURN UNIVERSITY
SAMUEL GINN COLLEGE OF ENGINEERING

30

Acknowledgements

- NSF Collaborative Research (REU Supplement): Processing and Properties of Cellulose Films for MEMS Applications (CMMI-1131633)
- Davis Lab group at Auburn University

31

AUBURN
UNIVERSITY
SAMUEL GINN
COLLEGE OF ENGINEERING

Questions?

?

9

AUBURN
UNIVERSITY
SAMUEL GINN
COLLEGE OF ENGINEERING

32

Birefringence: Interference Colors

- Michel-Levy chart can be used to determine the birefringence

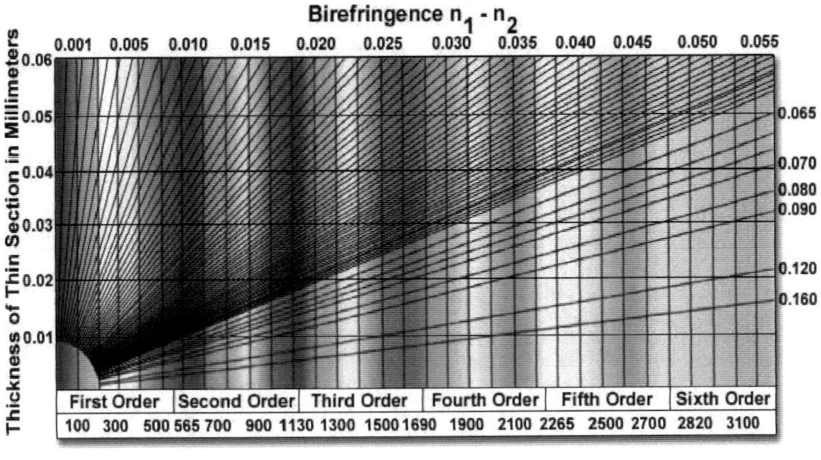

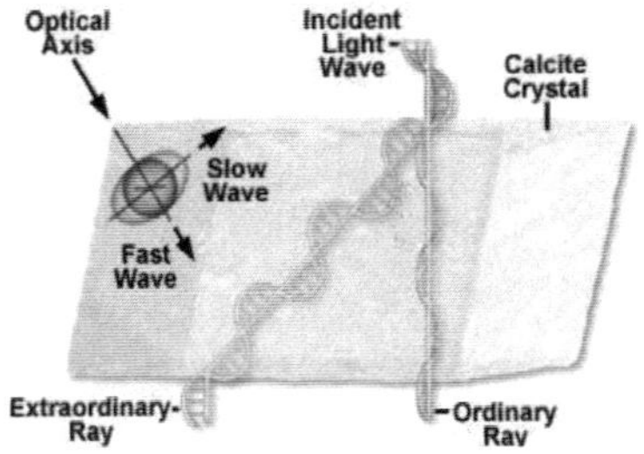

33 Images from: http://www.microscopyu.com/articles/polarized/birefringenceintro.html

AUBURN UNIVERSITY
SAMUEL GINN COLLEGE OF ENGINEERING

429540 Surface Modification of Cellulose Nanofibrils By Surfactant-Free Emulsion

Abstract:

Hydrophobic surface of cellulose nanofibrils (CNFs) was successfully prepared using surfactant-free emulsion polymerization of acrylate monomers. The polymerization reaction proceeded satisfactorily using persulfate initiator and the polymers synthesized in the process can encapsulate the surface of CNFs. Several different surface-modified CNFs were carried out using different amounts and compositions of polymer. The products were characterized based on thermal and hydrophobic properties using TGA and contact angle analyzer. Current result indicates the feasibility of CNFs surface modification using water as aqueous medium as a cost-effective method.

416927 A Quantitative Determination of the Specific and Nonspecific Interaction Forces Between Carbohydrate-Binding Module of CBH1/Cel7A and Lignocellulosic Biomass

Abstract:

Introduction

Enzymatic hydrolysis of lignocellulosic biomass has been considered as one of the key steps in the biochemical conversion of cellulosic biomass to sustainable biofuels and biochemicals. One of the major inhibitory mechanisms preventing efficient enzymatic hydrolysis is the nonproductive binding of cellulase to lignin. It has been suggested that cellulases adsorb to lignin and cellulose *via* hydrophobic, hydrogen bonding and ionic bond interactions. However, the exact mechanism of cellulase binding to cellulose and lignin remain unknown. Thus, elucidating the mechanism behind cellulase adsorption to both cellulose and lignin is required for the development of effective enzymes or pretreatment technologies for biofuel production. To understand the specific effect of lignin on cellulase adsorption, a set of reference substrates derived from kraft, sulfite and organosolv pulping was utilized to identify the types of forces involved in enzymatic interactions with cellulosic materials. The reference substrates used have known surface composition of cellulose and lignin.

Most of fungal cellulases have two modules namely a catalytic domain (CD) module and a carbohydrate-binding module (CBM). It is believed that the CBM plays a role in increasing the concentration of CD on the substrate surface and thus increasing the enzymatic action on the insoluble substrate surface. Therefore, studying direct interaction mechanism between the lignocellulose surface and CBM allows us to elucidate the nonproductive binding component of cellulases to lignin.

Atomic force microscopy (AFM) has the unique ability to quantify interactions between two surfaces at the molecular level and in native environments. We utilized AFM to measure the total forces acting between CBM and reference substrates in buffered medium. By performing adhesion force measurements between CBM and substrate surfaces, the binding affinity of CBM to lignin and cellulose was quantified. In addition, the effect of lignin type on CBM binding affinity was discussed. Poisson statistical analysis model was then applied to the data to decouple the adhesion force into specific and nonspecific components. Based on the quantitative comparison of specific and non-specific forces between CBM and substrate surfaces, the dominant type of forces effective in enzymatic binding to cellulose and lignin were determined.

Materials and Methods

CBM of *Trichoderma reesei* Cellobiohydrolase I (Cel7A) was expressed in *Escherichia coli* (*E. coli)* BL21 (DE3) cells with ketosteroid isomerase (KSI) tag using the pET31b(+) expression vector. The histidine-tagged fusion protein was purified on Ni-NTA agarose column with 10-250 mM imidazole gradient. CBM peptide was cleaved from KSI using cyanogen bromide. CBM was further purified from KSI after cleavage using reversed-phase chromatography with 0-60% acetonitrile gradient in 0.1% trifluoroacetic acid (TFA). The presence of CBM in the eluent was confirmed by mass spectroscopy analysis. The fraction containing CBM was lyophilized and resuspended in Tris-HCl (10mM, pH 7.5, 150 mM NaCl) buffer.

Modified organosolv, sulfite, and kraft pulping were used to produce reference substrates from poplar. The chemical composition of the reference substrates was analyzed *via* Technical Association of the Pulp and Paper Industry (TAPPI) test methods (T236, T204, T222, T211, and T249). Surface lignin coverage was characterized by X-ray photoelectron spectroscopy (XPS). Prior to force measurements, fibers of reference lignocellulosic substrates were attached to silicon wafers coated with poly-L-lysine (PLL). AFM cantilevers were functionalized with CBM and interaction forces between substrate surfaces and CBM were collected in 50 mM sodium acetate buffer at pH 4.8 under ambient temperature and humidity. A 5 x 5 μm^2 area of substrate surface was scanned at a 1 Hz rate. Spatially collected force-distance retraction curves were processed using the Nanoscope Analysis 1.5 software to determine the number of adhesion events and the magnitudes of adhesion forces measured between CBM and substrates' surfaces. Poisson statistical model was applied to adhesion force data to decouple the total interaction force into specific (hydrogen bonding) and nonspecific (Lifshitz-van der Waals and hydrophobic forces) components.

Results and Discussion

Our results showed that both hydrogen bonding and hydrophobic interactions contribute to enzymatic binding to lignocellulosic substrate surfaces. As the surface lignin coverage increases on the substrates, the overall adhesion forces of CBM to substrate surface increases in a linear manner (Figure 1). The slope of the linear line for kraft pulped (KP) substrates is 1.5 folds higher than that for the (sulfite pulped) SP substrates. Although we don't have a third point for the organosolv pulped (OP) substrates, based on the behavior of two data points obtained for the OP substrates, the trend observed for the KP and SP substrates is most likely valid for the OP substrates (Figure 1) . This suggests that the effect of kraft lignin on increasing the binding forces of CBM is higher than the effect of lignosulfonates and organosolv lignin on such interactions. Lignin-rich substrates showed higher binding strength to CBM in comparison to lignin-free substrates. This suggests that the affinity of CBM to lignin is higher than cellulose.

In addition to overall adhesion forces, the decoupled specific and nonspecific forces were compared. As the surface lignin coverage increases, the nonspecific forces increased for all types of pulped substrates. However, substrates with kraft lignin had 1.3 and 1.4 folds higher nonspecific forces than the substrates with lignosulfonates and organosolv lignin, respectively. This suggests that in the presence of hydrophobic lignin (kraft lignin) on substrates' surfaces, hydrophobic interactions contribute significantly to the overall interactions with CBM in comparison to other types of lignins investigated. When the specific and nonspecific forces were compared, it was found that the specific forces are higher than the nonspecific forces. The specific forces of lignin-rich substrates derived from kraft pulping, sulfite pulping and organosolv pulping were 1.3, 1.6 and 1.4 folds higher than the nonspecific forces of the same substrates, respectively. Having 1.6 fold greater specific forces for the lignosulfonates suggest that hydrogen bonding interactions are more dominant in comparison to kraft lignin and organosolv lignin. Additionally, as the surface lignin coverage increases, the specific forces increase. This suggests that hydrogen bonding interactions involved in nonproductive binding of CBM to lignin. In summary, both hydrogen bonding and hydrophobic interactions contribute to enzymatic binding to lignocellulosic substrate surfaces. However, the chemical structure of the lignin available on the substrate surface determines the type of interactions that dominate how lignocellulosic surfaces interact with other surfaces of interest.

Conclusion

Understanding the types of interaction forces that dominate the binding of enzymes to the biomass surface or specific chemical groups will allow protein engineers to design new enzymes that can specifically target cellulose groups in the biomass. In addition, by applying specific pretreatment technologies capable of reducing the hydrophobicity of lignin and reducing the number of chemical groups in lignin structure that are involved in hydrogen bonding, nonspecific binding of cellulase enzyme to lignin can be minimized.

Acknowledgment

This work is supported by National Science Foundation (NSF), Grant number: 1067012. We gratefully acknowledge Dr. Dmitri Tolkatchev and Prof. Alla Kostyukova from the Gene and Linda Voiland School of Chemical Engineering and Bioengineering at Washington State University, for their valuable help in CBM expression and purification.

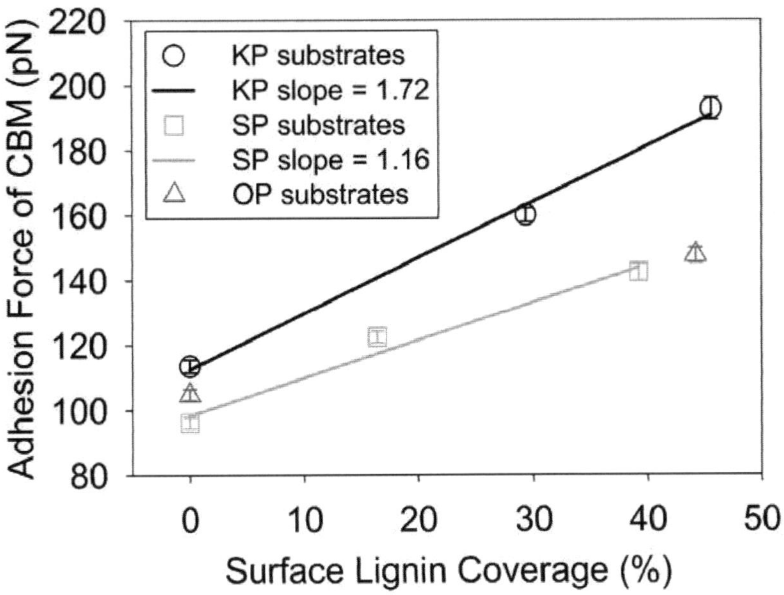

Figure 1. Adhesion forces of CBM to reference substrates as a function of surface lignin coverage. R^2 of linear regression lines are 0.99 and 0.97 for KP and SP substrates, respectively. Errors bars indicate the standard error of the mean.

425661 Zeta Potential of Lignin Plays a Positive Role in the Enzymatic Hydrolysis of Woody Biomass

Abstract:

The zeta potential related electrostatic interactions play a critical role in enzymatic hydrolysis of lignocellulsoic biomass. The objective of our study was to elucidate the positive effects of organosolv lignins on enzymatic hydrolysis. We observed the positive effects of organosolv lignins (EOL-CW, EOL-SG, EOL-BW and EOL-AS) from cottonwood, sweetgum, black willow and aspen) on enzymatic hydrolysis. On the contrary, we observed the negative effects of EOL-EH and EOL-LP from Eucalyptus and Loblolly pine on enzymatic hydrolysis. The addition of EOL-CW, EOL-BW and EOL-AS improved the 72 h hydrolysis yields of Avicel from 61% to 68%, 65% and 63%, respectively. The addition of EOL-EH and EOL-LP decreased the 72 h hydrolysis yields of Avicel to 53% and 56%, respectively. A good correlation between the zeta potentials of lignin and the hydrolysis yields indicated that the positive effects of organosolv lignins on enzymatic hydrolysis were governed by the zeta potential. The different *zeta potential of lignin* probably was related to the surface charges and hydrophobicity of lignin, which were further controlled by the physical and chemical properties of lignin. Our study suggested that organosolv lignins with high zeta potentials (>10 mV) could enhance enzymatic hydrolysis.

430966 Solvent Effects on the Dissolution and Precipitation of Cellulosic Biomass from Ionic Liquids

Abstract:

Chemical processing of biomass remains a challenge as the rigid inter- and intra- molecular hydrogen bonding network of cellulose renders it insoluble in nearly all aqueous and organic solvents. Alternatively, select ionic liquids (ILs) are capable of dissolving significant quantities. For instance, a promising IL, 1-ethyl-3-methylimidazolium diethyl phosphate [EMIm][DEP] is capable of dissolving more than 13 wt.% cellulose at only 60 ^oC. While pure ILs have been primarily targeted for biomass pretreatment, understanding the role of mixed solvent systems is important to improve momentum transport (i.e. lower mixture viscosity) and reduce the quantity of expensive IL required; two common criticisms of ionic liquid based processes. The vast majority of molecular solvents investigated dramatically decrease cellulose solubility in ILs and thus most are considered “antisolvents”. However, we have recently discovered that select quantities of polar aprotic solvents, when mixed with ILs, (despite having negligible solubility on their own) are capable of enhancing the thermodynamic solubility limit of cellulose in IL mixtures, even beyond the measured cellulose solubility limits of pure ILs. Furthermore, these “cosolvents” enhance transport properties of IL/biomass mixtures and reduce sensitivity to moisture (H_2O) which is known to severely impact dissolution capacities of cellulose in ILs. This presentation will illustrate solid-liquid phase equilibrium results for the dissolution of cellulose in various IL/cosolvent, IL/antisolvent, and IL/mixed solvent systems with the ionic liquid [EMIm][DEP]. Spectroscopic techniques including analysis by Kamlet Taft methodology will elucidate molecular interactions between ionic liquid and solvent species and provide an understanding of the solvation sphere around the IL in relation to cellulose dissolution within the various IL-solvent mixtures. Finally, rheological studies will demonstrate solvent effects on momentum transport. Our preliminary findings indicate that mixed IL/cosolvent systems could be even better solvents for cellulose dissolution than pure ILs from thermodynamic, transport, and economic perspectives. These results will be presented and discussed.

423208 Investigations into the Structural and Genetic Factors behind the Biomass Recalcitrance of Douglas Fir

Abstract:

Abstract Submission to Be Considered As an Oral Presentation at AIChE 2015 Annual Meeting in the Recalcitrance of Woody Biomass Session

Investigations into the Structural and Genetic Factors Behind the Biomass Recalcitrance of Douglas Fir.

Scott Geleynse, Xiao Zhang

Washington State University

Douglas Fir is a particularly challenging feedstock to utilize in the biochemical conversion to fuels and value-added chemicals due to its high levels of biomass recalcitrance. Earlier efforts have developed a foundation of data on the variation and genetics associated with this recalcitrance, along with a collection of tree samples screened for differing levels of recalcitrance. Despite their similar origin, some samples tested exhibited fermentable carbohydrate yields of over three times difference. In this presentation, we will discuss various investigations into the chemical and structural variations in several of these samples, along with a comparison of some genetic parameters to help shed light on the subtle differences that contribute to large variations in biomass recalcitrance of softwood biomass.

Presenting author:

Scott Geleynse

Bioproducts Sciences and Engineering Laboratory

School of Chemical Engineering and Bioengineering, Washington State University
2710 Crimson Way, Richland, WA 99354

Contact: scott.geleynse@email.wsu.edu, x.zhang@wsu.edu

425430 Dissolution of Cellulosic Fibers: Effect of Fiber Morphology and Size

Abstract:

Dissolution of cellulose, a nanostructured polymer abundant in nature, is a critical step for the efficient utilization of this renewable resource as a starting material for the synthesis of high value-added functional polymers and chemicals and also for biofuel production. The recalcitrance of cellulose microfibrils provides the major barrier to solubility of cellulose. There are a few solvent systems effective for direct dissolution of cellulose, and they operate under rather strict conditions of composition and temperature. In general, cellulose dissolution depends on its supramolecular and hierarchical structure, physico-chemical properties of solvent, and temperature. We investigate here the effect of morphology and size of cotton fiber, as the most important natural textile and a highly crystalline form of cellulose, on the mechanism of fiber swelling and dissolution. To this end, we have developed a phenomenological model for describing the swelling and dissolution of solid cellulosic fibers. The model considers cellulosic fibers as non-uniform crystalline polymers with different morphological properties among their cross sections due to the different supramolecular and morphological structures of cell-walls. The model parameters, i.e., the solvent effectiveness in decrystallization of cellulose, and the ability of solvent in cellulose chain untangling, was obtained by fitting to experimental data. The modeling parameters were then used to examine the influence of fiber size and structure (e.g., degree of polymerization and crystallinity) on the mechanism of dissolution process as well as on the total time required for decrystallization and dissolution of the fiber. The insights obtained from this model should facilitate the design of efficient solvent systems and processing conditions for woody biomass dissolution.

423294 Comparing Maximum Overall Pretreatment and Enzymatic Hydrolysis Sugar Yields from Poplar Wood Following THF Co-Solvent Enhanced Lignocellulosic Fractionation (CELF), Dilute Acid, and Organosolv Pretreatments

Abstract:

Because lignocellulosic ethanol production typically requires a costly pretreatment step to overcome biomass recalcitrance, advanced pretreatments could lead to significant reductions in overall ethanol production costs. The University of California at Riverside recently invented a novel pretreatment we now call Co-solvent Enhanced Lignocellulosic Fractionation (CELF) pretreatment that uses tetrahydrofuran (THF) in a homogeneous solution with water containing dilute sulfuric acid to achieve very high recovery of hemicellulose sugars in solution and removal of lignin during the CELF process. Furthermore, the highly glucan enriched solids produced by CELF can be readily digested to high glucose yields with much lower and more economically attractive enzyme loadings than typical for most pretreatments. This study optimized CELF pretreatment of poplar wood to maximize total xylose and glucose yields from the combined operations of pretreatment and subsequent enzymatic hydrolysis and compared the results to maximum total sugar yields from dilute acid and ethanol organosolv pretreatments of the same substrate. CELF pretreatment conditions that achieved the highest combined total sugar yields from poplar wood at high enzyme loadings were determined to be 160°C for 15 minutes with 0.5% H_2SO_4 as the catalyst at a 1:1 THF/H_2O ratio. At these conditions, CELF removed more than 90% of the lignin from poplar and realized xylose plus glucose yields of more than 95% at low enzyme dosage. Yield data will be presented over a range of enzyme loadings from 2 to 60 mg/g for CELF pretreatment compared to optimized application of dilute acid and ethanol organosolv pretreatments to the same poplar material. In addition, composition analysis, SEM imaging, and other characterization tools will be employed to identify key differences in pretreated solids features that could be responsible for differences in sugar yields, particularly at lower enzyme loadings.

421932 Polyoxometalates As Catalyst for Valorization of Lignin to Produce Aromatic Chemicals and Hydrogen

Abstract:

Lignin is the second abundant component of lignocellulosic biomass after cellulose. With its rich aromatic structure and huge production, it has potential to be the major renewable aromatic chemicals and hydrogen resource. Polyoxometalates (POMs) are a class of promising catalysts which can not only convert the lignin into chemicals, but also easily be regenerated. In this study, we obtained a quantitative yield of the product vanillin at mild conditions. Furthermore, when the POM regenerated at the anode in the newly designed hydrogen-producing electrochemical cell, hydrogen can be produced at cathode with near 100% Faradaic efficiency and very low electricity consumption.

433204 Investigation of Nanostructural Characteristics of Lignin-ABA Block Copolymer Aggregates Leading to Use of Micelles As Nanoreactors for Lignin Conversion

Abstract:

Nanostructural self-assembly of lignin derivatives incorporated in polyethylene oxide-*block*-polypropylene oxide-*block*-polyethylene oxide (PEO-*b*-PPO-*b*-PEO) micelles in aqueous alkaline solution has been investigated. Hypotheses for copolymer micelle incorporation of lignaceous species include the "hydrophobic-hydrophilic interaction model" in which lignin derivatives with apolar-polar characteristics, known to exist in the effluent of alkaline pretreatment of various lignocellulosic materials, have an affinity for the micelles, and tend to form relatively organized structures within them. Here we show how the concentration and chemical/structural features of lignin monomers and small molecule lignin models determine their solubilization (guest) behavior in nanoscopic micelles.

Using (^{1}H) NMR, we have found that increasing concentrations (0.01-0.2 mmol) of lignin-based model compounds guaiacol, eugenol, and phenol in Pluronic® F68-D_2O solutions results in significant decreases of the PEO/PPO blocks' proton chemical shifts, and at a critical concentration, dramatic shifts due to gross structural transitions in the micelles. Although lignin also affects copolymer chemical shifts significantly in these micellar systems, large structural transitions seen in the models are seemingly not available to lignin-micellelar structures for several reasons. We present evidence that polymer micelle-guest interactions depend on both chemical functional group characteristics of guests (i.e. polarity, H-bonding ability) and pi-orbital stacking in ordered aromatic/conjugated groups.

Our results demonstrate how the loci of incorporated guests in A-B-A block-co-polymer are affected by these features. The current study is a starting point for utilizing micelles as a nanoreactor platform for converting partially-depolymerized lignin biopolymer to value-added monomers such as the antioxidants guaiacol and eugenol.

433319 Lightweight Functional Material from Sustainable Resources: Lignin-Soy Aerogels

Abstract:

Recent development in functional materials using natural polymers has triggered a lot of interest in applications that required both lightweight and mechanical strength. Lignin and soy protein have been used in different applications in composite materials. However, the synergy between these two materials have not been fully explored in such venue. In this work we are reporting on in the production of lignin – soy protein aerogels. The effect of different lignin and soy protein ratios, coadjutant polymers (polyethylene oxide, polyvinyl alcohol and cellulose nanofibrils), as well as, the addition of a cross-linker (glutaraldehyde) on the morphology, rheological and mechanical properties were explored. After freeze drying low density, highly porous aerogels were obtained. The cross-linker increases the mechanical properties of the aerogel. Overall results indicate the synergy between lignin and soy proteins in the fabrication of aerogels and shows promise for the development of low-cost functional materials.

433656 Generating Ultrapure Lignins from Kraft Pulp Mill Lignins Via the ALPHA Technique

Abstract:

Lignins can now be recovered from paper mill Kraft black liquors by several processes that have recently become commercial or are under advanced development. Two of these processes are (1) LignoBoost and (2) SLRP (Sequential Lignin Recovery and Purification). The "Kraft lignins" obtained from these processes have ash contents of 1-2% and total metals contents of several thousand ppm. Of interest to us is the isolation of lignins that are greatly reduced in metals and ash content and could serve as biorenewable precursors for materials such as carbon fibers and composites.

We have developed a technique for purifying Kraft lignins that can drive the metals content down to very low levels (<100ppm) in a few simple steps. We call the method ALPHA (Aqueous Lignin Purification Using Hot Acids). Hot acetic acid-water mixtures that partition the lignin between an acetic-acid-rich liquid phase and a water-rich liquid phase is the key feature that makes this technique work. ALPHA has been successfully applied to both LignoBoost and SLRP lignins. The technique requires only minor modification to the basic process, depending on the raw material source, in order to obtain the desired ultrapure (<100 ppm total metals) lignins (UPLs). In conclusion, ALPHA has shown itself to be a robust technique for producing UPLs from different lignin sources encompassing different purities.

434000 Vanillyl Alcohol: A Renewable Epoxy Resin Building Block

Abstract:

Polymers derived from renewable resources are becoming considerably attractive as sustainable alternatives to their petroleum-derived counterparts. Lignin is a natural and abundant raw material made up of cross-linked aromatics that could potentially be broken down to yield renewable bio-based monomers and resins for applications in high-performance thermosets and composites. In this work, vanillyl alcohol was utilized as a promising lignin-based building block for the synthesis of epoxy monomers and resins. Specifically, diglycyl ether of vanillyl alcohol (DGEVA) was synthesized and its chemical structure was confirmed by ^{1}H-NMR, ^{13}C-NMR, FT-IR, and GPC. Bimodal resin blends of diglycidyl ether of bisphenol A (DGEBA) with varying weight ratios of DGEVA as well as bimodal resins blends of DGEBA with other single aromatic epoxies, diglycidyl ether of hydroquinone and diglyical ether of 4-hydroxymethyl phenol, were prepared and cured with a standard diamine, methylenebis(cyclohexylamine), to investigate their processing-structure-property relationships. Additionally, DGEVA was blended and cured with a (hemi)cellulose-based diamine to investigate the viability of a highly bio-based cured epoxy resin. We present the thermogravimetric and thermomechanical properties as well as the extents of cure that were obtained and analyzed to gain insight into the behavior of the cured resins. Moreover, we present green chemistry metrics results of the renewable epoxy resin systems.

434763 Systems Biology-Guided Biodesign of Consolidated Lignin Conversion

Abstract:

Lignin is the second most abundant biopolymer on the earth. However, due to the recalcitrant nature of the complex polyphenolic structure, the conversion of lignin remains a major challenge for sustainable biorefineries. We hereby elucidated the lignin degradation mechanisms and carbon utilization pattern in a unique lignin-utilizing *Pseudomonas putida* strain (A514) with comparative genomics and systems biology. The mechanistic study further guided the design of three functional modules to validate the molecular mechanisms and enable a consolidated lignin bioconversion route. First, *P. putida* A514 coordinately up-regulated a dye peroxidase-based enzymatic system for lignin depolymerization. The augmentation of this system by production and secretion of a heterologous, highly active dye peroxidase significantly improved the cell growth on Kraft lignin. Together with the NMR analysis of lignin structure, the study established the role of bacterial peroxidase in effective degradation of lignin, instead of model compounds, to promote cell growth. Second, the lignin substrate induced production of a network of enzymes and pathways for aromatic compound degradation, which can be enhanced by over-expressing genes encoding key enzymes. Third, the β-oxidation of fatty acid was up-regulated when A514 was grown on lignin and vanillic acid. Based on the carbon utilization pattern, the functional module for PHA production was designed to channel β-oxidation products for increasing PHA yield. The integration of the three functional modules enabled the direct production of PHA from lignin polymer, elucidated the lignin conversion mechanisms in bacteria, and laid out the concept for designing a consolidated lignin conversion route.

439901 Functionalization of Pyrolyzed Biomass and Their Characterizations

Abstract:

Carbonaceous materials are commonly applied as an effective reinforcement in polymer composites. However, current materials such as carbon fibre and carbon nanotubes can be prohibitively expensive for widespread application, and are often derived from non-renewable resources. Pyrolyzed biomass (biochar) could provide an alternative carbon-rich reinforcement material that is derived from renewable resources. In addition to its high carbon content, biochar is a low-cost material that can be produced from a variety of agricultural feedstocks. In this study, biochar from pyrolyzed perennial grass was studied as a potential reinforcing material. The tailoring of biochar surfaces was investigated using oxidative acid treatments to potentially improve adhesion to certain polymer materials. The treated biochars were characterized using various methods to determine their chemical, thermal, morphological and crystalline characteristics. The study revealed that biochar is a promising candidate for composite materials, as the oxidative treatments were effective in improving the surface functionality of the material. FT-IR, EDS and elemental analysis revealed significant changes in chemical structure following oxidative acid treatment. Such modified biochar can be investigated further for their suitability in polymer composites uses.

Acknowledgements: The authors are thankful to the Ontario Ministry of Agriculture, Food and Rural Affairs (OMAFRA) – University of Guelph "Product Development and Enhancement through Value Chains Research Theme" (Project # 200399); OMAFRA- University of Guelph "Bioeconomy for Industrial Uses Research Theme" (Project # 030055) and the Natural Sciences and Engineering Research Council (NSERC), Canada Discovery Grants (Project # 400322) for their financial support.

437215 Effect of Stirring Ball Milling on Structure and Hydrolysis of Camphorwood Sawdust

Abstract:

Mechanical activation can destroy the crystal structure of cellulose and promote its degradation. Camphorwood sawdust was mechanically activated with a stirring ball mill to study its change of structure and hydrolyzation catalyzed by CO_2 in hot water. The effects of milling time, mass ratio of ball to biomass, ball grinding rotation speed on the crystal structure, particle size distribution, surface topography and hydrolysis of biomass were studied. The structural changes and particle size distribution of camphor wood chips were characterized by scanning electron-microscopy (SEM), X-ray diffraction (XRD), Fourier transform infrared (FTIR) spectroscopy and laser granularity analyzer. The results show that stirring ball mill can remarkably reduce the particle size of camphorwood sawdust and make the crystal structure of sawdust cellulose destroyed at the ball grinding rotation speed 450rpm, milling time 2h and mass ratio of ball to biomass 30:1. The crystallinity declines from 60.9% to 21.4%. D50 decreases from 25.07μm decreased to 9.29μm,and the corresponding specific surface area increases from 1.79 $m^2 \cdot g^{-1}$ to 4.88 $m^2 \cdot g^{-1}$. After activation, hydrolysis was conducted at temperature of 180 °C, pressure of 8 MPa CO_2 and reaction time of 60 minutes. Cellulose conversion reaches 37.75%, which is nearly four times of that of unactivated raw biomass. Reducing sugar content in hydrolysate was determined by the dinitrosalicylic acid (DNS) method. The concentration of reducing sugar content in the hydrolysate is 6.02 g/L after activation, which increases by 31.15% compared to unactivated raw material.

HYDROTHERMAL CARBONIZATION (HTC) OF DIFFERENT BIOMASS TYPES

Joan G. Lynam*, M. Toufiq Reza, Wei Yan, Victor R. Vasquez, Charles J. Coronella,

Chemical & Materials Engineering Dept.

University of Nevada, Reno

AIChE Annual Meeting
Monday, November 9, 2015

*Presenting

Biomass as Fuel

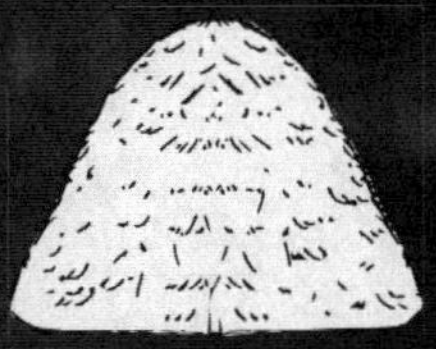

- Complex, expensive logistics
  - Seasonal availability
  - Widely distributed
  - Low fuel density, low mass density – expensive to ship, high bulk
- Unique challenges:
  - Poor storage stability
  - Expensive to mill – want more friable product
  - Diverse handling characteristics of bioma

Hydrothermal Carbonization of Different Biomass Types JG Lynam AIChE Annual Meeting November 9, 2015

To Pretreat Biomass Using Hydrothermal Carbonization

- Hot compressed water (200 – 280 °C)
- Short contact time (< 30 min)
- High Pressure : vapor pressure of liquid water (1- 5 MPa)

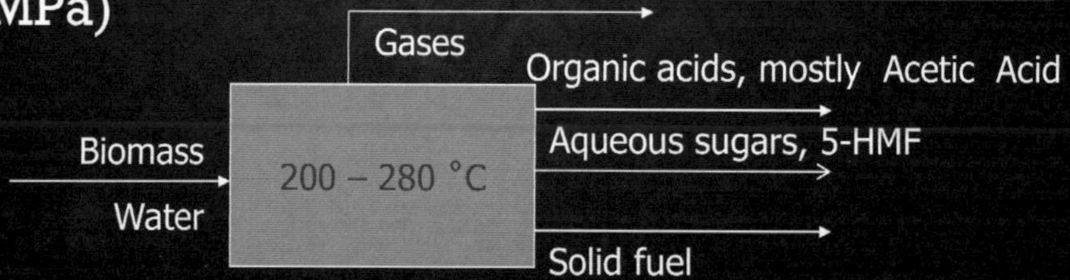

Hydrothermal Carbonization of Different Biomass Types JG Lynam AIChE Annual Meeting November 9, 2015

Initial Experimental Design

- High-pressure 100-mL Parr reactor
- Temperature: 200 °C or 260 °C
- Batch experiments- 8 g biomass immersed in liquid water
- 5:1 or 20:1 water:biomass (w:w)
- Reaction time: 5 min or 20 min
- Biomass particle size 0.16 mm or 0.24 mm diameter
- Initially Loblolly pine, then feedstock varied

Hydrothermal Carbonization of Different Biomass Types JG Lynam AIChE Annual Meeting November 9, 2015

HTC Hydrochar

- Product is soft, friable solid
- Odor of "sugary charcoal" or "burnt marshmellow"

12

Hydrothermal Carbonization of Different Biomass Types JG Lynam AIChE Annual Meeting November 9, 2015

Results Loblolly Pine

- No Effect from Water : Biomass ratio or Particle Size variation
- Increase in Reaction Time Decreases Mass Yield Slightly

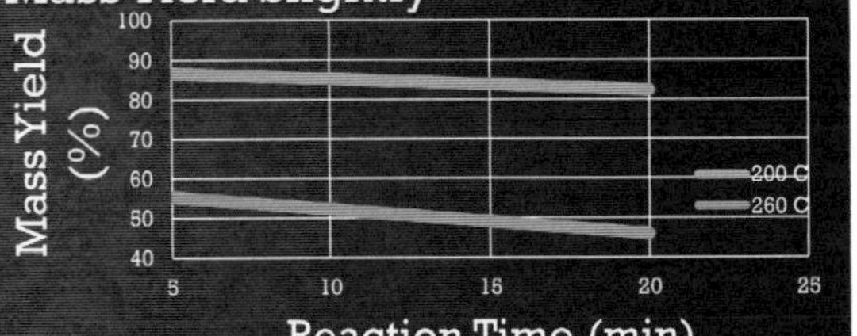

- Main Effect from Temperature Increase

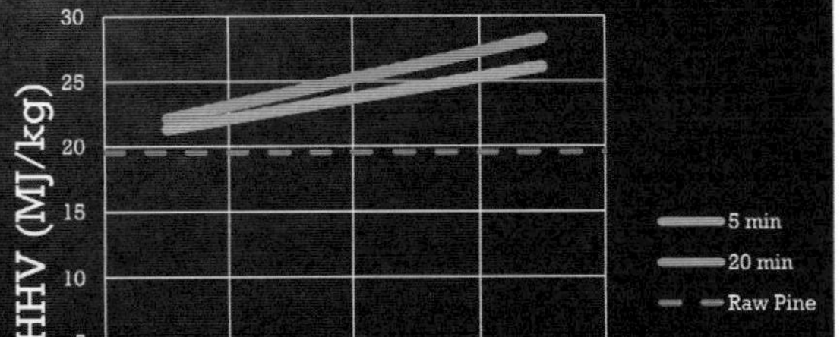

Hydrothermal Carbonization of Different Biomass Types JG Lynam AIChE Annual Meeting November 9, 2015

Hydrophobicity

Loblolly Pine	Moisture from proximate analysis
Raw	3.6
200 °C	1.3
260 °C	1.3

HTC increases Hydrophobicity, making biomass easier to dry

HTC worked with Loblolly Pine to Increase HHV: Why?

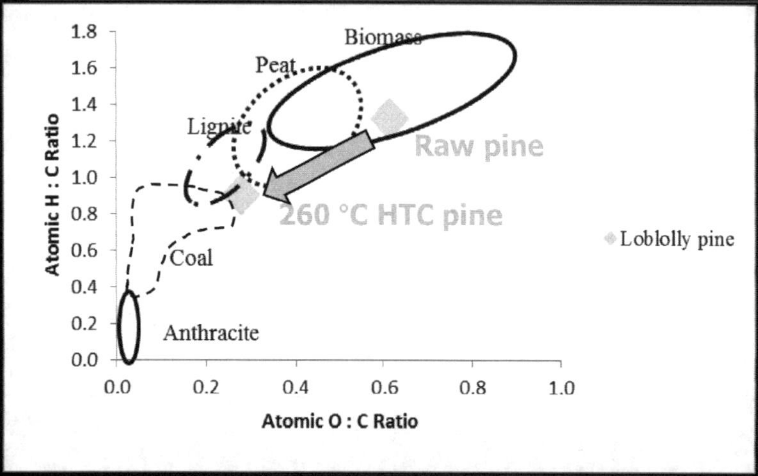

- Dehydration and Decarboxylation are reducing the amount of oxygen in the biomass, which increases its higher heating value

Van Krevelen Diagram

What about other biomass and temperatures?

- Corn Stover - a primary agricultural residue
- Rice Hulls - a secondary agricultural residue
- Tahoe Mix (Jeffrey Pine And White Fir) - a forest product removed to reduce wildfires
- Switch Grass - a primary energy crop
- Loblolly Pine - a primary forest crop

Try HTC at 200, 230, and 260 °C, 5 min, water:biomass ratio of 5:1

Hydrothermal Carbonization of Different Biomass Types JG Lynam AIChE Annual Meeting November 9, 2015

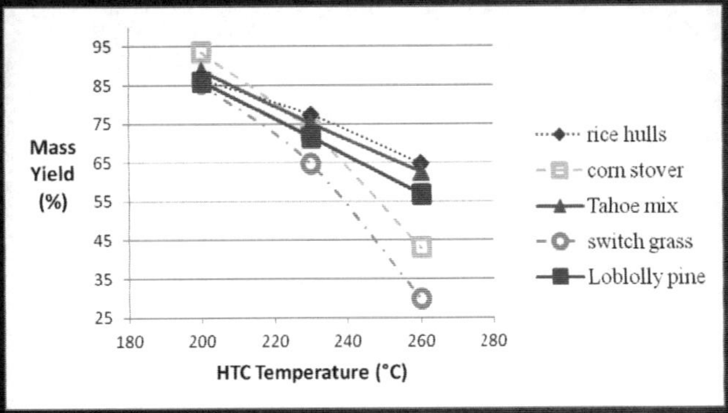

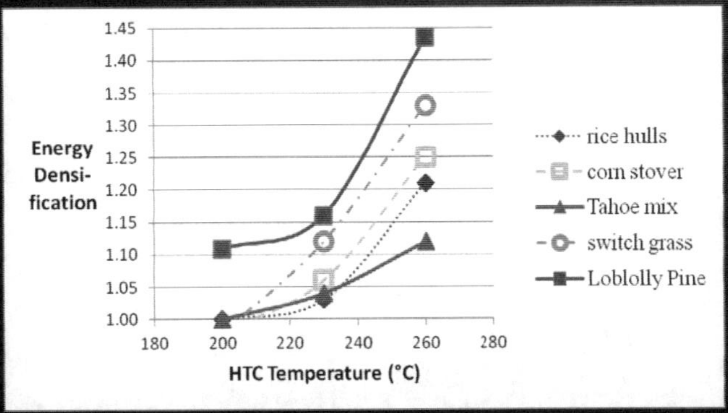

Results

- **Grassy biomass gives low mass yield**
- **All biomass has lower mass yield & higher energy densification with increasing reaction temperature**

$$Energy\ densification = \frac{HHV\ hydrochar}{HHV\ raw\ biomass}$$

Hydrothermal Carbonization of Different Biomass Types JG Lynam AIChE Annual Meeting November 9, 2015

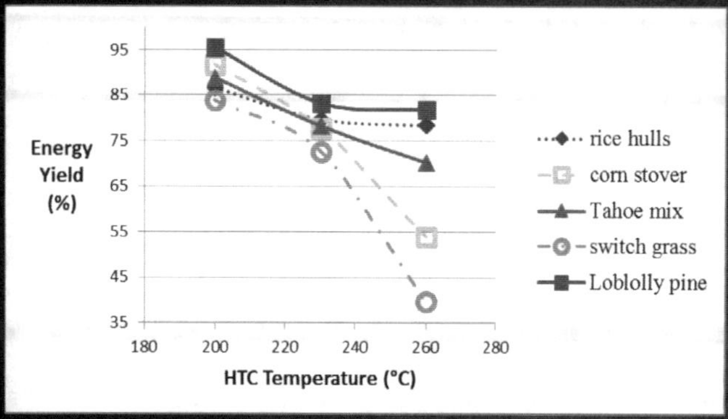

Results

- **Grassy biomass gives low mass yield, so they give low energy yield**

$$\textbf{Energy yield} = \textit{mass yield} * \textit{energy densification}$$

Why do we get these kind of results?

Hydrothermal Carbonization of Different Biomass Types JG Lynam AIChE Annual Meeting November 9, 2015

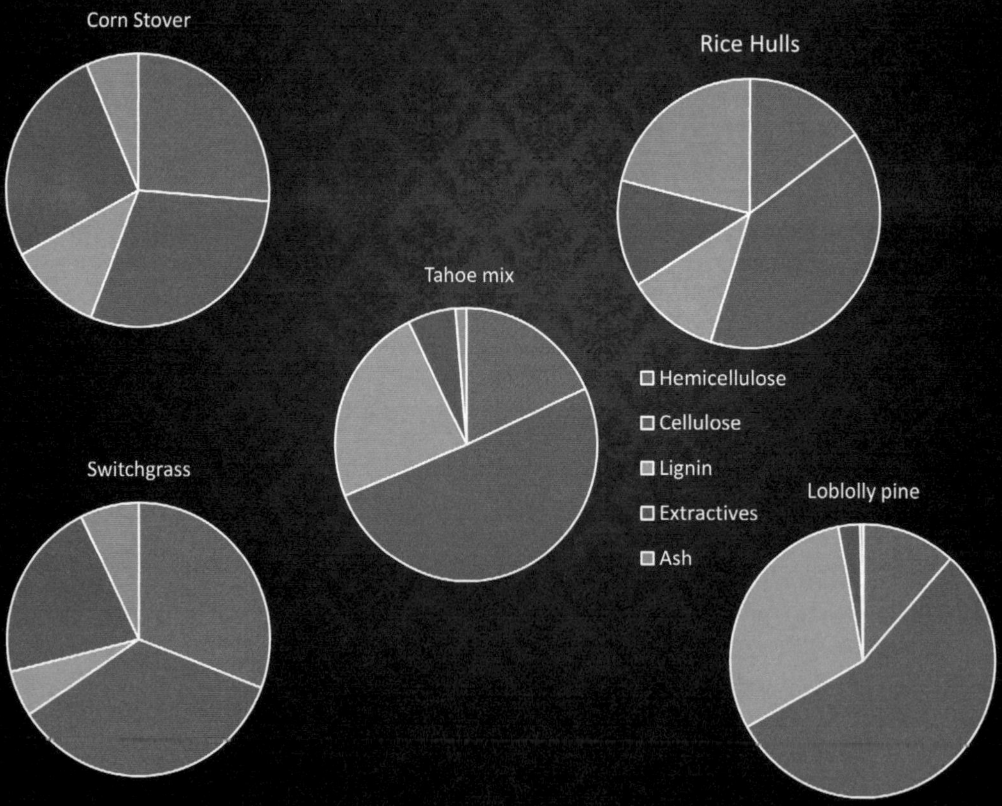

Hydrothermal Carbonization of Different Biomass Types JG Lynam AIChE Annual Meeting November 9, 2015

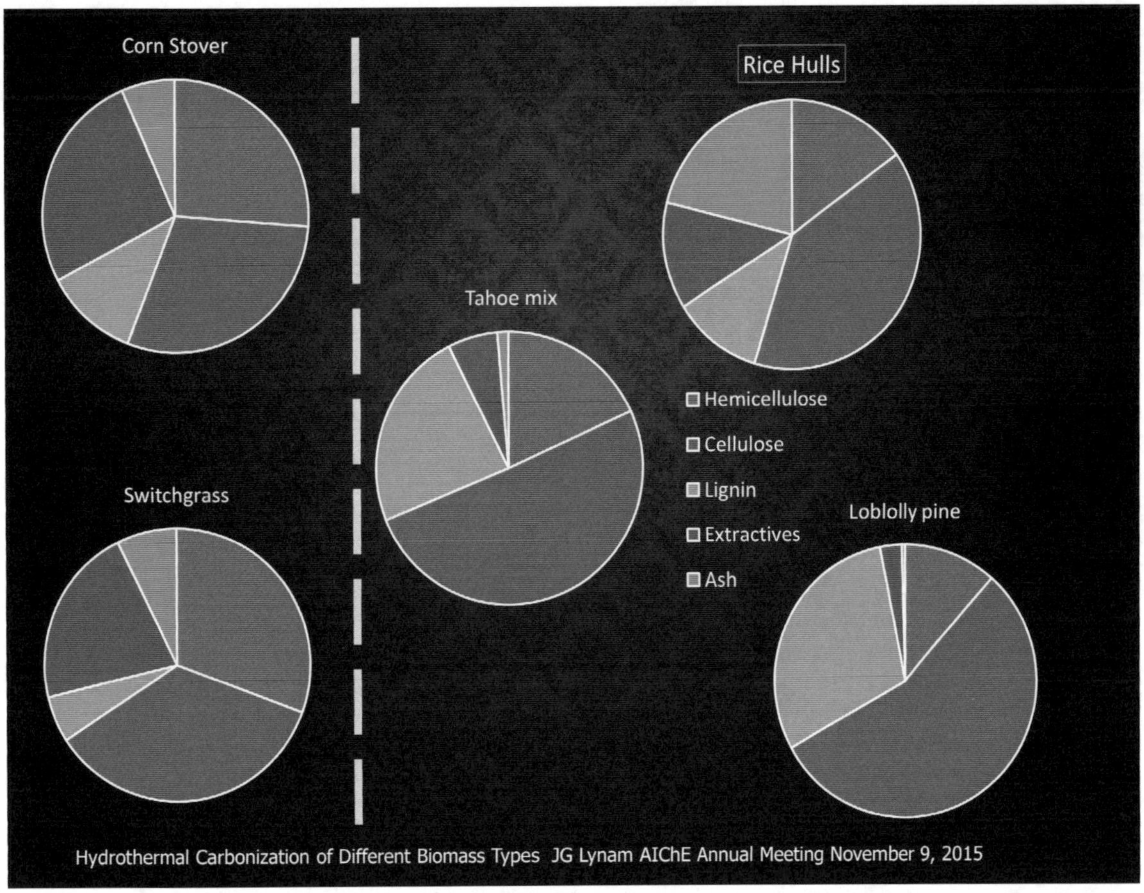
Corn Stover
Rice Hulls
Tahoe mix
Switchgrass
Hemicellulose
Cellulose
Lignin
Extractives
Ash
Loblolly pine
Hydrothermal Carbonization of Different Biomass Types JG Lynam AIChE Annual Meeting November 9, 2015

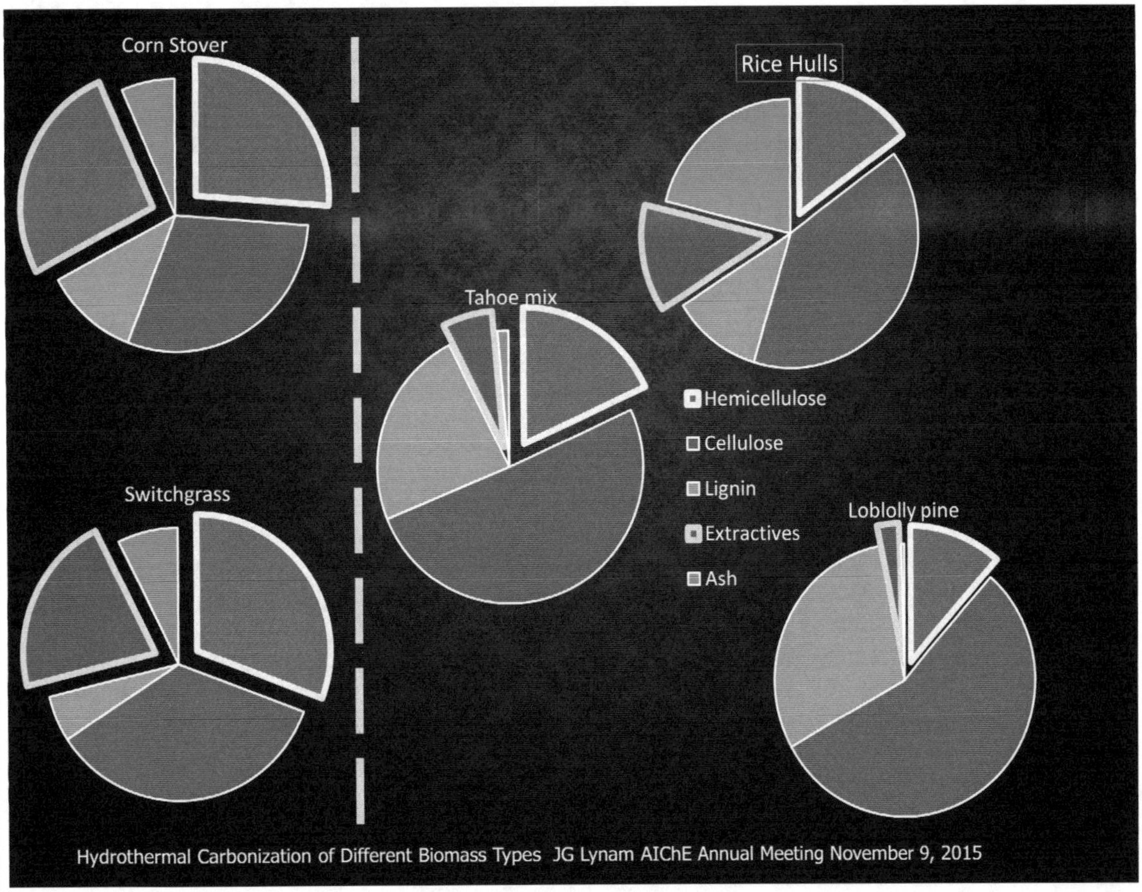
Corn Stover
Rice Hulls
Tahoe mix
Hemicellulose
Cellulose
Lignin
Extractives
Ash
Switchgrass
Loblolly pine
Hydrothermal Carbonization of Different Biomass Types JG Lynam AIChE Annual Meeting November 9, 2015

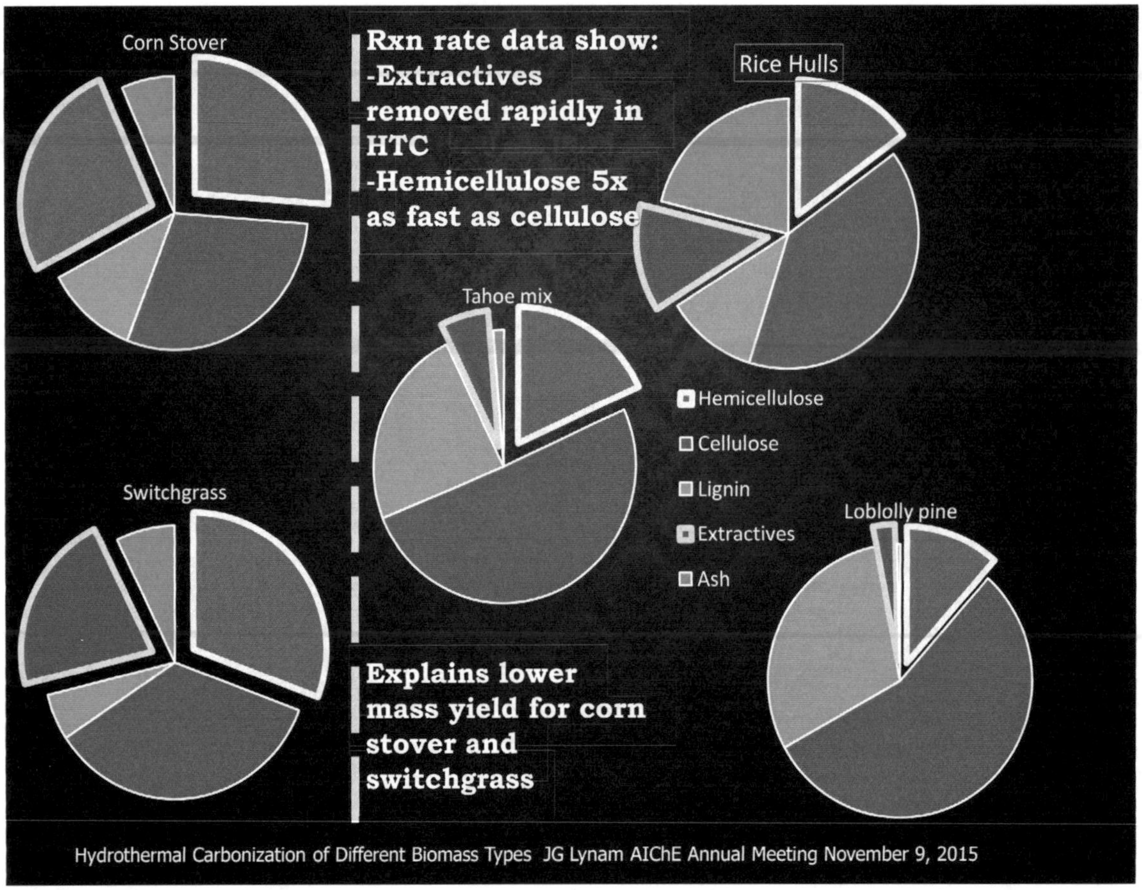
Corn Stover
Rxn rate data show:
-Extractives removed rapidly in HTC
-Hemicellulose 5x as fast as cellulose
Rice Hulls
Tahoe mix
Hemicellulose
Cellulose
Lignin
Extractives
Ash
Switchgrass
Loblolly pine
Explains lower mass yield for corn stover and switchgrass
Hydrothermal Carbonization of Different Biomass Types JG Lynam AIChE Annual Meeting November 9, 2015

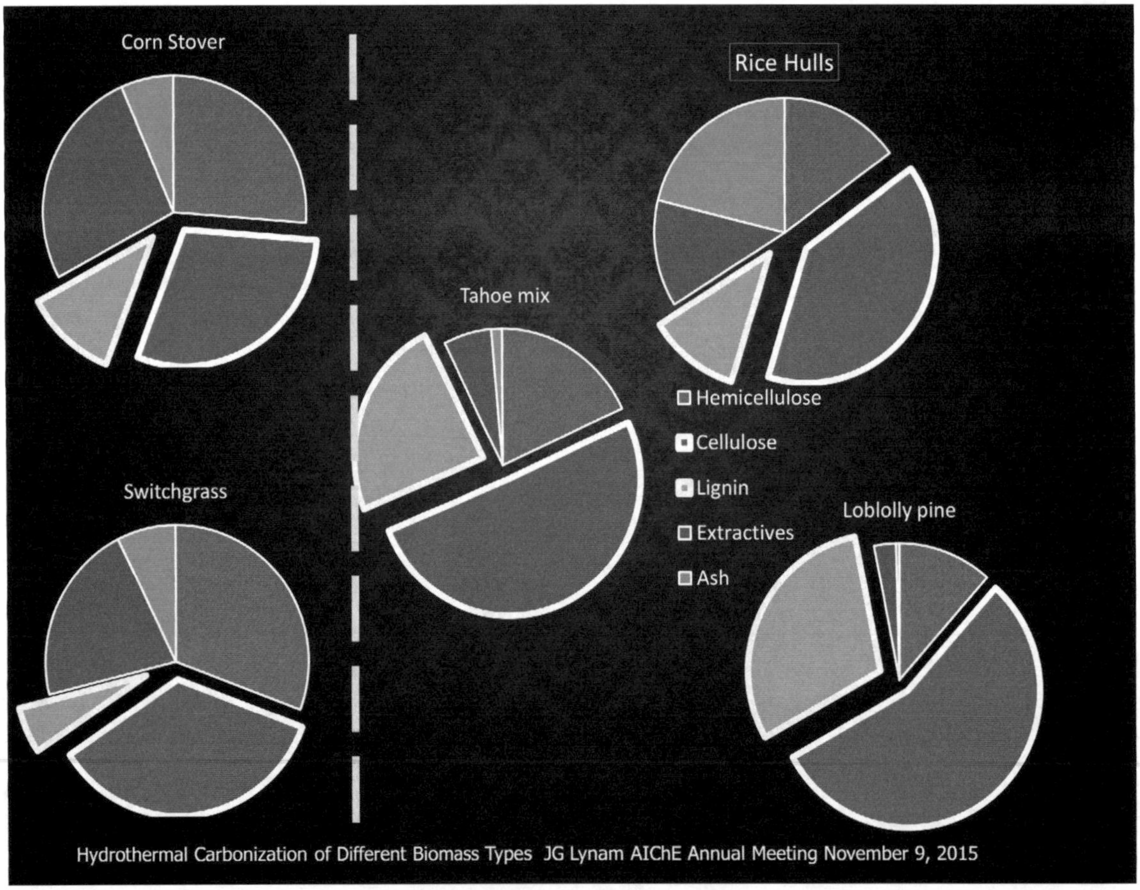
Corn Stover
Rice Hulls
Tahoe mix
Hemicellulose
Cellulose
Lignin
Extractives
Ash
Switchgrass
Loblolly pine
Hydrothermal Carbonization of Different Biomass Types JG Lynam AIChE Annual Meeting November 9, 2015

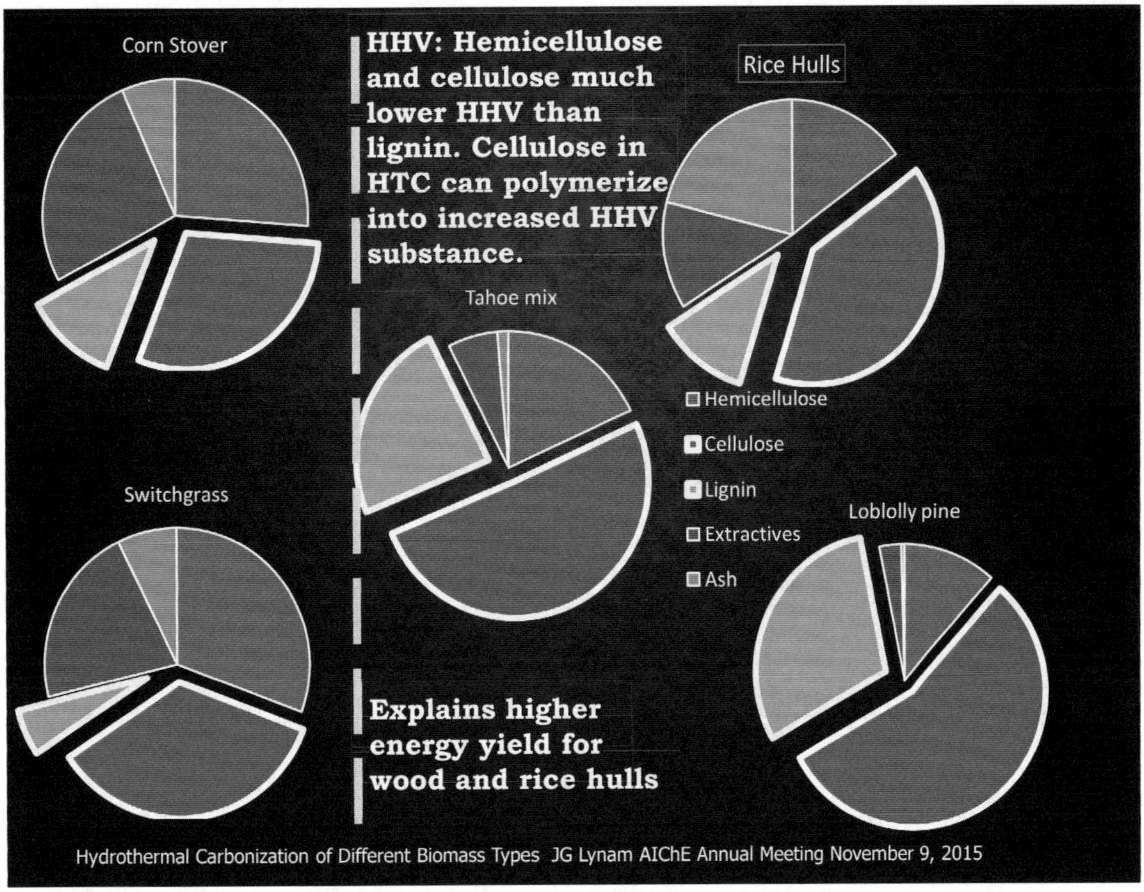
Corn Stover
HHV: Hemicellulose and cellulose much lower HHV than lignin. Cellulose in HTC can polymerize into increased HHV substance.
Rice Hulls
Tahoe mix
Hemicellulose
Cellulose
Lignin
Extractives
Ash
Switchgrass
Loblolly pine
Explains higher energy yield for wood and rice hulls
Hydrothermal Carbonization of Different Biomass Types JG Lynam AIChE Annual Meeting November 9, 2015

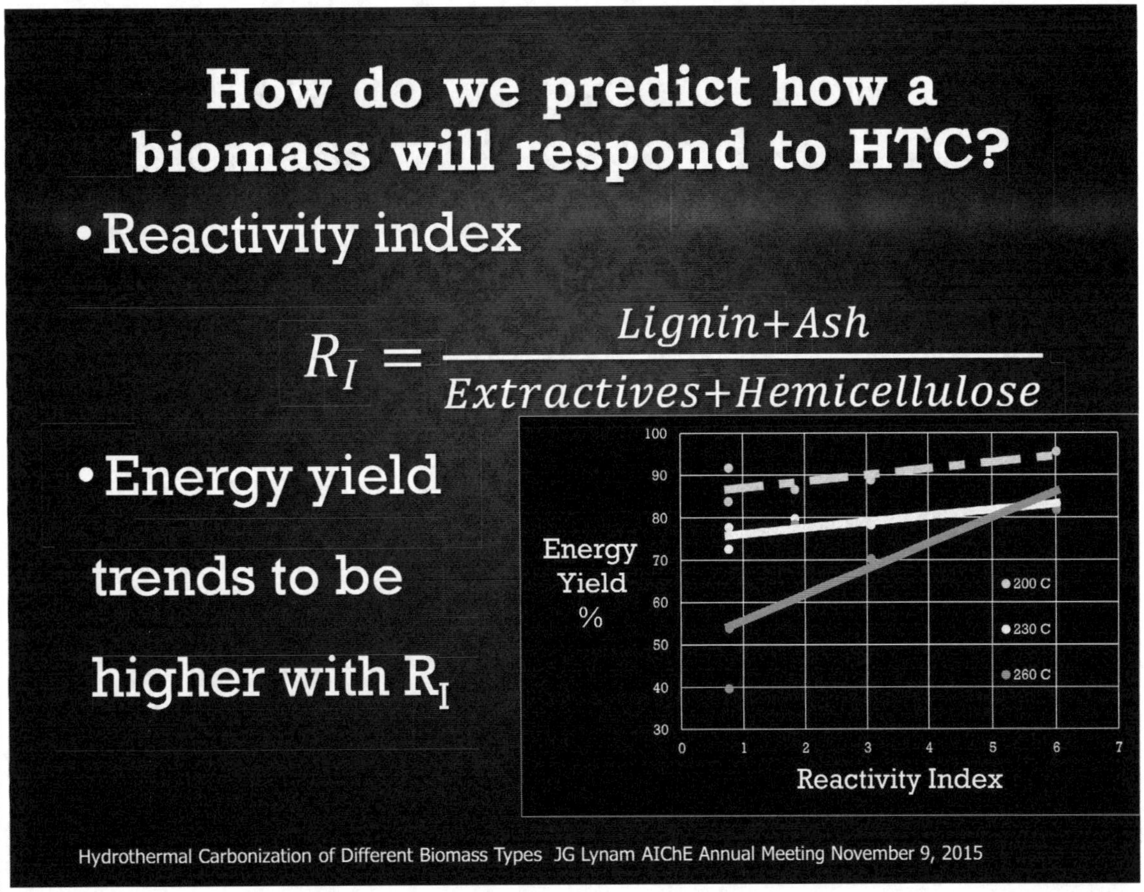
How do we predict how a biomass will respond to HTC?
• Reactivity index
$R_I = \frac{Lignin+Ash}{Extractives+Hemicellulose}$
• Energy yield trends to be higher with R_I
Energy Yield %
100
90
80
70
60
50
40
30
0
1
2
3
4
5
6
7
Reactivity Index
200 C
230 C
260 C
Hydrothermal Carbonization of Different Biomass Types JG Lynam AIChE Annual Meeting November 9, 2015

SEM of Hydrochar

- Loblolly Pine

Raw:

– relatively smooth surface

– extractives fill in structure

HTC at 200:

- hemicellulose solubilized

- lignin unwrapping?

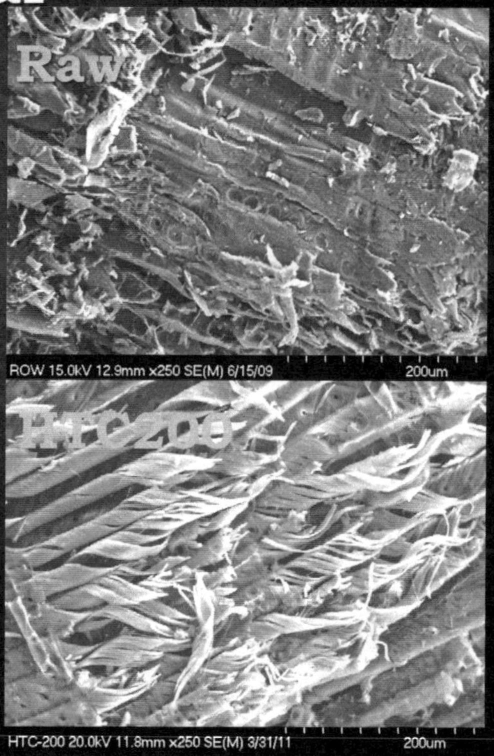

Producing a Higher Density Solid Fuel from Biomass

Dry Torrefaction:

- Advantages
  - Ambient Pressure process
- Disadvantages
  - Drying first required
  - Reaction time > 30 minutes- several hours
  - Less Higher Heating Value (HHV) than HTC pretreated product

Hydrothermal Carbonization:

- Advantages
  - Short reaction time
  - No drying required
  - More hydrophobic – easier to dry afterward
  - Increased HHV
  - Easy to pelletize
- Disadvantages
  - Higher pressure operation
  - Water recycling required

Conclusions

- Hydrothermal carbonization (HTC) can be used to produce a hydrophobic solid fuel in 5 minutes of reaction time without initial drying from various biomass feedstocks
- HTC can improve energy density by up to 43%
- Reaction temperature is the significant process variable
- Grass-like biomass feedstocks of higher hemicellulose and extractive content produced low mass yield
- Softwood biomass of higher lignin and cellulose content produced a higher mass and energy yield
- A reactive index ratio of the raw biomass' (lignin and ash)/(hemicellulose and extractive) can predict energy yield

Hydrothermal Carbonization of Different Biomass Types JG Lynam AIChE Annual Meeting November 9, 2015

Acknowledgements

- This material is based upon work supported by the National Institute of Food and Agriculture, U.S. Department of Agriculture, under Agreement No. 2013-67011-21011 and the Department of Energy (Contract Number: DE-FG36-01GO11082).

USDA NIFA United States Department of Agriculture National Institute of Food and Agriculture

- The authors acknowledge meaningful conversations with project partners, including Larry Felix of the Gas Technology Institute, Kent Hoekman of the Desert Research Institute, and Craig Einfeldt of Changing World Technologies. Experimental work of Mr. Tapas Acharjee and Mr. Jason Hastings is gratefully acknowledged.

Any opinions, findings, conclusions, or recommendations expressed in this publication are those of the author(s) and do not necessarily reflect the view of the U.S. Department of Agriculture.

Hydrothermal Carbonization of Different Biomass Types JG Lynam AIChE Annual Meeting November 9, 2015

Acknowledgements

- This material is based upon work supported by the National Institute of Food and Agriculture, U.S. Department of Agriculture, under Agreement No. 2013-67011-21011 and the Department of Energy (Contract Number: DE-FG36-01GO11082).

USDA NIFA United States Department of Agriculture National Institute of Food and Agriculture

- Joan G. Lynam, M. Toufiq Reza, Wei Yan, Charles J. Coronella*, Victor R. Vásquez, *Hydrothermal carbonization of various lignocellulosic biomass*. Biomass Conversion Biorefinery 2015. Volume **5**(2): 173-181.

Thank you! Questions?

Hydrothermal Carbonization of Different Biomass Types JG Lynam AIChE Annual Meeting November 9, 2015

422202 The Relation of Lignin Structural Properties and the Conversion Reactivity to Value-Added Chemicals

Abstract:

Abstract Submission to Be Considered as an Oral Presentation at ***AICHE 2015Annual Meeting*** in *T4A03 Chemical Conversion Processes in Forest/Plant Biorefineries*

The Relation of Lignin Structural Properties and the Conversion Reactivity to Value-added Chemicals

Ruoshui Ma[1], Mond Guo[1], Vincent Hebert[1], Jinwen Zhang[1], Xiao Zhang[1]

1. Washington State University, Richland, WA, 99354

Lignin is the largest nature source of renewable material with an aromatic skeleton. Converting lignin to low molecular weight aromatic and phenolic compounds (LMWPC), dicarboxylic acids (DCA), and quinones offers attractive opportunities to produce a range of high value products/chemicals.

In this presentation, we will report our current progress on characterization of several representative biorefinery lignins and investigating the relation between their structural properties and their conversion reactivity towards the value-added platform chemicals (e.g. LMWPC, DCA).

Presenting author: Ruoshui Ma, Chemical Engineering, Washington State University, Richland, WA, 99354

Contact: ruoshui.ma@email.wsu.edu

434892 Integrated Thermochemical Process Design for Co-Producing Liquid Fuels and Propylene

Abstract:

Biorenewable value-added products are regarded as promising near-term solutions to increasing environmental requirements and strong dependence on imported petroleum. During the past two decades, biochemical, thermochemical, and catalytic routes for biomass conversion have gained significant attention. For the thermochemical routes, gasification and fast pyrolysis are two potential technologies that offer high efficiencies for the production of liquid transportation fuels and chemicals [1,2]. The gasification method first converts biomass to syngas which will be further upgraded towards liquid fuels/chemicals via Fischer-Tropsch synthesis or methanol synthesis, while the fast pyrolysis method converts biomass to bio-oils which are then converted to gasoline/diesel through hydrotreating and hydrocracking.

In this presentation we systematically design and optimize an integrated gasification-fast pyrolysis based biorefinery process. In addition to the production of liquid transportation fuels including gasoline and diesel, propylene is also considered as a co-product. In the designed process, the heat for the gasifier and fast pyrolysis reactor will be provided by bio-char combustion.

The hydrocracking unit is replaced by a catalytic cracking unit [3] which will process the long-chain hydrocarbons from the bio-oil hydrotreating unit and wax from the Fischer-Tropsch product fractionator. C_{5+} gasoline, distillate, propylene and coke are the main products of the catalytic cracking unit. Hydrogen for the bio-oil hydrotreater, naphtha hydrotreater, distillate hydrotreater, and C_4 isomerizer can be provided through pressure swing adsorption of synthesis gas from the gasification unit or provided by the reforming of water soluble pyrolysis oil. All the generated CO_2 will be captured for re-use or sequestration. The developed process superstructure, which considers multiple catalysts and reactor types for certain processing units, is formulated as a mixed integer nonlinear programming model. To reduce the computational complexity of the MINLP, simple surrogate models for the gasifier are introduced. The bilinear, trilinear, quadrilinear, and concave terms are reformulated for enhancing the MINLP model to be solved to global optimality. The economics associated with the designed integrated process and the stand-alone gasification-FT /fast-pyrolysis are compared. Trade-offs between process economics and hydrogen source selection are considered. The optimal topology and the corresponding optimal operating conditions are evaluated under different scenarios, for example, maximum propylene production and maximum liquid fuels production. The designed integrated process exhibits high flexibility to cope with uncertainties on market demands and feedstock sources.

Reference:

[1]. Wright MW *et al*. Techno-economic analysis of biomass fast pyrolysis to transportation fuels. *Fuel*. 2010; 89: S2-S10.

[2]. Swanson RM *et al*. Techno-economic analysis of biomass-to-liquids production based on gasification. *Fuel*. 2010; 89: S11-S19.

[3]. Komvokis VG et al. Upgrading of Fischer-Tropsch synthesis bio-waxes via catalytic cracking: Effect of acidity, porosity and metal modification of zeolitic and mesoporous aluminosilicate catalysts. *Catalyssis Today*. 2012; 196: 42-55.

426386 Graphitic Bio-Carbon from Lignin Biomass Synthesized with Nickel Nitrate Catalyst

Abstract:

Abstract

Lignin is a high volume byproduct of pulp production that is currently used as an energy supply. In this current work, graphitic porous carbon materials synthesized from lignin in the presence of nickel nitrate as a catalyst. The synthesis scheme of graphitic carbon consists of i) lignin was subjected to a hydrothermal treatment in a unique sonication reactor at 275 °C around and 1400 psi to form bio-char, ii) the bio-char was impregnated with nickel nitrate and carbonized at temperature ranged from 800 to 1100 °C under an inert atmosphere, iii) the resulting graphitic carbon was washed with 10% (w/w) HCl in order to remove the residual nickel particles. The characterization and degree graphization of biomass changed with temperature. Structural and surface properties of the resulting graphite-based materials were investigated by FTIR, BET surface analyzer and elemental analysis. Thermal gravimetric analysis (TGA) was performed to determine the thermal resistance of the materials. XRD results exhibited that the highest degree of crystallinity of graphitic carbon synthesized with NiNO3 catalysis obtained at 1100 °C degree. Electrical transport properties of particles were characterized using two-terminal current-voltage (I-V) curves and shows very high conductivity for NiGC1100 Nickel catalyzed graphitic carbon. The as-prepared material has many potential applications such as an electrode in supercapacitor and Li-ion battery. This application is currently under testing.

422786 High Efficient Biomass-to-Hydrogen Conversion By Polyoxometalate Solution Catalyzed Electrolysis

Abstract:

Hereby we present a novel method for electrolysis that can directly convert natural polymeric biomasses, such as starch, cellulose, lignin, and even switchgrass and wood powders to hydrogen. Polyoxomatelate (POM) was used as phot- or thermal- catalyst. POM first form a charge transfer complex with the biomass by absorbing light or heat. The charge transfer complex was then release electrons on the carbon anode under very low electric field, and simultaneously release H^+. As a result, the H^+ diffuse to cathode and was reduced to hydrogen. Comparing to the reported electrolysis of biomass for hydrogen production, our method could save significant electric energy.

432915 Qualitative and Quantitative Characterization of Biomass Using FT-IR Microspectroscopy and X-Ray Diffraction

Abstract:

The ability to quickly and accurately determine key quality characteristics of biomass prior to processing in a thermochemical biorefinery is critical in determining the types of products, and co-products, which will be produced in the conversion process. Proximate (moisture, ash, fixed carbon, volatiles) and ultimate (C, H, N, O, S) analyses and chemical composition (e.g., cellulose, hemicellulose, lignin) result in quality parameters of importance to thermochemical conversions. For example, oxygen content causes low heating value and corrosiveness, while cellulose, hemicellulose, and lignin each convert to unique products during thermochemical conversion for both pyrolysis and hydrothermal liquefaction. Rapid screening models based upon FT-NIR and FT-IR spectroscopy are being developed to rapidly predict fixed C, H, N, O, volatiles, and total ash of select feedstocks to avoid the long analysis times and high costs of current methods.

In addition, FT-IR microscopy offers the potential to analyze biomass for proximate, ultimate, and chemical composition at spatial scales as small as 5 μm. [1] In this work, models based upon FT-NIR and FT-IR are developed and validated to predict fixed C, H, N, O, volatiles, and total ash in switchgrass samples, as well as glucan, xylan, and lignin for a set of mixed feedstock types. These models are then applied to FT-IR microscopy data to predict the local variability of specific properties at scales of 100 μm or smaller. A focal plane array (FPA) detector with a pixel resolution less than 5 μm is used to rapidly create chemical composition maps of select regions of the sample surface. This technique promises new methods to investigate the chemical composition of biomass materials at the cell and tissue level as well as track how preprocessing operations impact those properties at the microscale.

Finally, biomass ash structure has been probed for various types of woody biomass, at various stages of pyrolytic conversion, using x-ray diffraction (XRD). Investigation of the biomass ash structure at various stages of pyrolysis allows for determination of the structural changes that could take place during processing at a thermochemical biorefinery. Knowledge of the inorganic content of the biomass will enhance the ability to determine useful co-products from what is now considered waste material, such as the use of biomass ash in the production of cement based materials.[2]

1 Allison, G. G. *et al.* Measurement of key compositional parameters in two species of energy grass by Fourier transform infrared spectroscopy. *Bioresource technology* **100**, 6428-6433 (2009).

2 Rajamma, R. *et al.* Characterisation and use of biomass fly ash in cement-based materials. *Journal of hazardous materials* **172**, 1049-1060 (2009).

430162 Value-Added Products from Thermal Treatment of Biomass Pyrolysis Oil

Abstract:

Pyrolysis of lignocellulosic biomass produces bio-char, gaseous byproducts, and liquid fractions rich in phenolic, acidic, and oxygen-rich compounds, otherwise known as pyrolysis oil. Utilization of pyrolysis oils is hampered by a high water content (as much as 30 wt%), as well as instability and corrosiveness due to oxygenated species (aldehydes, phenolics, and carboxylic acids). We investigated thermal treatment of pyrolysis oils as a means to produce a value-added product with potential uses in soil amendment and chemical adsorption applications. Products and byproducts were characterized analytically (GC-MS, BET, SEM), and a mass balance was evaluated for the process. We concluded that oil from biomass pyrolysis could be upgraded using moderate reaction conditions to generate a product useful for non-fuel applications

421378 Corncob Hydrolysate Based Media for Pigments Production By Penicillium Purpurogenum GH2: Kinetics and Modeling

Abstract:

Fungi provide a readily available alternative source of natural pigments with potential industrial applications due to the wide range of colors that are able to produce. Despite of their major properties (antioxidant, antibacterial and cytotoxic activities), the industrial application of natural pigments is still uncompetitive to synthetic colorants due to their production costs. Efforts need to be addressed in developing a low cost bioprocess for the production of such natural fungal pigments by using inexpensive raw materials and a process improvement. Lignocellulosic materials have gained extensive attention as a promising alternative source for the industrial production of valuable compounds by fermentation. Selection of a suitable lignocellulosic waste will depend on the availability of the raw materials for any geographical zone and the bioprocess required. It has been reported that corncob hydrolysate contains a high level of xylose due to its high composition of hemicellulose and thus represent an abundant and inexpensive carbon source that can be converted to pigments by microorganisms.

This study aimed to determine the feasibility of using a corncob hydrolysate as an alternative substrate for pigment production by *Penicillium purpurogenum*GH2 that has been previously reported as a potential pigment producer strain.

To hydrolyse the sugars oligomers to the correspondent monomers a dilute sulphuric acid hydrolysis treatment was used (temperature: 120 °C, pressure: 15 psi, H_2SO_4: 1%, 90 min, solid-liquid ratio: 10% (w/w)). The hydrolysate was detoxified using activated charcoal. The detoxified liquor without any nutrient supplementation was diluted to obtain a xylose concentration of 15 g/L and used as fermentation medium (HM). A synthetic Czapeck-dox modified medium (CDM) previously optimized for the production of pigments by this strain containing (g/L): D-xylose 15.0, $NaNO_3$ 3.0, $MgSO_4 \cdot 7H_2O$ 0.5, $FeSO_4 \cdot 7H_2O$ 0.1, K_2HPO_4 1.0, KCl 1.0 and ethanol 20.0 was used as control. Submerged fermentation experiments were carried out at 200 rpm, 30 °C for 8 days.

In both media the microorganism utilized two carbon sources (xylose and ethanol in CMD and xylose and glucose in HM) for its growth and production of metabolites. For future process optimization using a hydrolysate-based media, it is important to fully understand how the microorganism is consuming each substrate and how this simultaneous uptake affects cell growth and pigment production. Therefore pigment production, growth and substrate consumption kinetics using CMD and HM were studied. Unstructured fermentation models were applied to obtain valuable information as parameters with clear biological meaning.

The Verlhust logistic model was applied to describe the carbohydrate consumption during fermentation. This model describes substrate consumption as function of the residual substrate (S_{res},g/L), the ratio between the initial volumetric rate of substrate consumption and the initial substrate concentration (S_0', h^{-1}) and the lag time (k, h) before the carbohydrate is consumed. Biomass growth (X, g/L) was expected to be function of the consumption of both carbon sources (S_i, g/L) in each medium (xylose and ethanol in CDM; xylose and glucose in HM) taking in account the yield ($Y_{x/si}$, g/g) of biomass formed by each substrate. Production of pigments was described using the Luedeking-Piret model. This model assumes that the production rate (dP/dt) depends on the instantaneous biomass concentration (X,g/L), the growth rate (dX/dt) and takes into account when the production is limited by the exhaustion of a carbon source defined as limiting substrate (S_L, g/L). It also contains 2 parameters with significant biological meaning, α for production related to growth and $\beta$for production related to maintenance.

The models described accurately the experimental kinetic data with correlation coefficients (R^2) of 0.97 and 0.99 for CDM and HM substrates respectively. Total pigment production obtained with HM was 6% lower that the pigments produced in the control CDM, however cell growth, production and substrate consumption presented different patterns in both media.

When CMD was used, the microorganism rapidly adapted to the medium showing a lag phase less than 24 h. Initial consumption rates differed from xylose (4.13 g/L.h) and ethanol (1.65 g/L.h) thus even though the obtained lag time for the utilization of ethanol was low (k=24 h), its assimilation started at a low rate. The microorganism used ethanol and xylose for growing and showed a typical diauxic growth due to the utilization of two carbon sources. The biomass yield obtained by ethanol ($Y_{X/S}$=0.11) was higher than with xylose ($Y_{X/S}$=0.08). Pigment production was related to grow with α=4.91 and β=0.

In HM medium, biomass growth presented an extended lag phase, indicating that the microorganism needed more time to adapt to the medium. The microorganism first used the most assailable and simple carbohydrate, glucose. The initial consumption rates varied from 0.38 g/L.h in glucose to 1.12 g/L.h in xylose. This confirmed the poor adaption of the microorganism to the hydrolysate. The time lag k for xylose assimilation was nearly 80 h. The highest yield of biomass per substrate was achieved by glucose ($Y_{X/S}$=0.53) meanwhile the biomass formed with xylose was almost negligible ($Y_{X/S}$=0.02) meaning that the microorganism did not used xylose to continue growing thus it did not show a diauxic type growth. Pigments production started after 80 h that corresponded to the lag time in xylose. Pigments production was associated with maintenance with values of α=0 and β=4.70.

Corncob hydrolysate was proved to be an efficient alternative medium for pigment production by *Penicillium purpurogenum* GH2. The hydrolysate without any nutrient supplementation showed high production yield of total pigment comparable to that obtained with the synthetic medium. The production of pigments was related to cell maintenance when the hydrolysate was used as fermentation media, thus for future process optimization the first approach would be the reduction of the extended initial lag phase.

425713 New Biopolyester from Co-Product of Biodiesel Industries: Synthesis, Characterization and Blending

Abstract:

Glycerol is the main co-product of biodiesel industries with a global production higher than 3 billion lbs that is expected to double by 2020 reaching nearly 6 billion lbs. Glycerol polyesters have been synthesized by previous researchers mainly for biomedical applications using pure glycerol with 99wt% of glycerol content. Technical glycerol, an industrial refined glycerol from biodiesel industries with 97% or higher glycerol content is a potential candidate for the synthesis of polyesters in applications where glycerol purity is not a crucial issue. In this work, technical glycerol based polyesters were synthesized by simple polycondensation procedures with diacids as comonomers and subsequently these polyesters were blended with several biodegradable plastic matrices in a extrusion procedure. The effect of the molar ratio of reactants, type of diacid employed and reactive extrusion mode in the mechanical properties of the biobased blends have been studied. It has been found that the thermoplastic materials produced with up to 20wt% of glycerol polyester in the blend have acceptable mechanical and rheological properties for injection molding applications.

Acknowledgements

The authors are thankful to the Ontario Ministry of Agriculture Food and Rural Affairs (OMAFRA)- University of Guelph Bioeconomy-Industrial uses Research Program Theme (Project # 200001); OMAFRA – New Directions & Alternative Renewable Fuels Research program (Project # SR9223); the Natural Sciences and Engineering Research Council (NSERC), Canada for the Discovery grants individual (Project # 400322); and the Ministry of Economic Development and Innovation (MEDI), Ontario Research Fund - Research Excellence Round - 4 program (Project # 050231 and 050289) for their financial support to carry out this research work. The first author acknowledges Chilean National Scholarship Program for Graduate studies from CONICYT-Chile for their financial support. The authors are also thankful to BIOX Corporation, Methes Energies International Ltd and Veenstra Farms, Canada, for providing crude glycerol samples.

437372 Model of Ink Particle Separation in Column Flotation Deinking

Abstract:

Column flotation is a novel technique of separation of hydrophobic particles from a suspension. An experimental investigation of column flotation as applied to de-inking of a 50-50 mix of photocopier and laser-print waste paper is described. Experiments were conducted in both the batch and the continuous mode of operation. Hand-sheets were made from the de-inked pulp and the brightness and ink particle size distributions were measured. The kinetic constants for de-inking were determined for the continuous modes. A comparison between the laboratory scale column flotation and other flotation deinking methods is evaluated and modeling of the process described.

433703 Pervaporation-Assisted Catalytic Conversion of Xylose to Furfural

Abstract:

Furfural produced from the biomass-derived xylose may serve as a platform molecule for sustainable fuel production. The Brønsted acid-catalyzed dehydration of xylose to furfural is plagued by side reactions that form a set of soluble and insoluble degradation products, collectively known as humins, which reduce the yield of furfural. The formation of humins can be minimized by removal of furfural, either by steam stripping or by liquid-liquid extraction (LLE). However, both these techniques are very costly. The goal of this study was to demonstrate the feasibility of using pervaporation, a membrane process, to remove furfural as it is produced. A laboratory-scale reactor/membrane system was designed, built, and tested for this purpose and its performance for furfural production was compared with that achieved by carrying out the reaction with and without furfural extraction by LLE. Furfural production assisted by pervaporation or LLE produced comparable amounts of furfural, and more than could be achieved by reaction without extraction. Our results showed that under our experimental conditions, pervaporation extracted a smaller fraction of the furfural produced than did LLE (72% vs 89%), but pervaporation produced a permeate phase with a furfural concentration 10 times greater than that present in the extractant phase obtained by LLE. It is noted that further improvement in the separation of furfural from the aqueous phase where it is produced can be achieved by operating the membrane at the reaction temperature and using a thinner pervaporation membrane. A 3-fold membrane-thickness reduction was simulated using MATLAB and was shown to increase the extracted fraction of furfural produced to 90% while maintaining the same furfural concentration in the permeate. An analogous change in LLE (*i.e.* increasing the organic:aqueous volume ratio) would increase the extracted fraction of furfural produced at the cost of diluting the furfural concentration in the extractant phase.

437285 Kinetics of the Enzymatic Hydrolysis of Papermaking Pulp Fines

Abstract:

Waste fibers from papermaking are often rejected due to quality and production issues. Since the rejected fibers also known as 'fines' are discharged as solid waste placing significant strain on landfills, bioconversion into sugars by hydrolysis of their cellulosic content can be very attractive to the paper industry. Earlier experimental investigations studying this hydrolysis have established the roles of the enzyme activity (given in terms of FPU), temperature and pH.

The present work develops a model for the kinetics of the enzymatic hydrolysis of cellulosic pulp fines based on the physico-chemical interactions. The cellulolytic enzymes adsorb to the outer surfaces of the fines after which they diffuse into the interior pores. The fine particles are assumed to be spherical, containing two distinct kinds of pores: one at the macro-scale and another at a micro-scale allowing slow progression of the enzymes into the solid cellulosic matrix. Enzymatic hydrolysis reactions are assumed to occur at the micro-scale within the cellulosic matrix. Convective mass transfer boundary layers reduce the enzyme transport across the external particle surfaces. The main utility of this model is in estimating the optimal reaction conditions and assist in scaling studies of these enzymatic hydrolysis processes.

410638 Bioreactor Process Model for Microalgae Production Systems

Abstract:

Microalgae are a promising feedstock for production of fuels and value-added chemicals. In order to assess the economic, energetic and environmental performance of algae based system in an industrial scale biorefinery, it is important to model algae productivity accurately. In this work, a first principle based growth model of microalgae biomass and lipid production is developed for an outdoor pond system. The primary factors that affect microalgae growth and accumulation of lipids, such as solar irradiance, temperature, specific characteristics of species, uptake rates of nutrients and carbon dioxide, are incorporated in the bioreactor process model. Year-round algae productivity potential is computed taking into the seasonal effects on microalgae growth. As there are significant uncertainties associated with the different parameters of the growth model, detailed sensitivity analyses are performed. Results show that solar irradiance and maximum specific growth rate are the most significant parameters in microalgae bulk growth. Results from the dynamic model can be used to optimize and scaling up of algae systems for efficient production of fuels and chemicals for variety of algae species as well as geospatial locations.

437184 Integrated Bioprocessing and Separation and Purification Technology for Advanced Biofuels and Biochemicals

Abstract:

In an effort to address the USA Congressional biofuels mandate of 36 billion gallons per year goal by 2022 and the production of advanced biofuels and biochemicals, it is imperative that we develop integrated technologies overcoming the barriers for advanced biofuels and chemicals. Butanol is not only an advanced drop in biofuels but also an important chemical and industrial solvent. It has several advantages over ethanol as a drop-in biofuel such as higher energy content, potential for higher blending percentage with gasoline, lower vapor pressure, and lower hygroscopy. Butanol can be produced by fermentation of cellulosic biomass sugars using Clostridium acetobutylicum or C. beijerinckii under anaerobic conditions. However, there are many unsolved challenges for making biobutanol technically, and economically viable. The unsolved challenges lie in severe product (butanol) inhibition during bioprocessing, which leads to low ABE (Acetone, Butanol and Ethanol) yield, low productivity, and very low product concentration (<3 wt%), causing high energy consumption and high costs in downstream product separation and purification. There are two ways for solving these problems. One is the modification of microorganisms for ABE fermentation by metabolic and genetic engineering, which could keep the microorganisms alive and active under higher concentration of products in the broth. This could significantly increase the product yield, productivity, and concentration and hence reduce the production costs. However, this is still an unrealized long term goal. Another approach is the development of efficient separation and purification processes for product recovery. And, even if the modification of microorganisms becomes a reality, product separation and purification will still remain a major critical challenge (Huang et al. 2014).

In this work, we develop integrated bioprocessing (fermentation) -in situ separation and purification (product removal) (IBSP) processes. Among the three most promising separation methods – solvent extraction, membrane pervaporation and adsorption, we studied membrane pervaporation integrated in-situ with bioprocessing by process modeling and simulation to determine the effect of integrated (IBSP) approach. The integration of biobuanol fermentation with in situ separation and purification technology can not only remove product inhibition and improve the yield and productivity but also reduce the energy consumption and hence reduce the production costs. This is also applicable to other bioproducts manufacture and can play a vital role in the success of the future bio-economy.

[Reference] Huang et al., Separation and purification of biobutanol during bioconversion of biomass. Separation and Purification Technology, 2014 (8): 132:513–540.

437347 Reuse of Deinking Sludge from Recycled Paper Industry in Light Fired Clay Bricks

Abstract:

The sludge generated from recycle paper industry and disposal chalanges associated with it are increasing with the growth of wastepaper based industry in India. The utilization of such wastes in concrete applications like building materials can not only reduce the waste but also generate revenue. This work presents a study carried out at laboratory scale for the use of recycle paper industry sludge in light weight bricks for some specific applicaion. The fired clay brick were made and tested with varied proportion of clay and sludge from recycle paper industry. The clay is replaced by 0%, 10%, 20%, and 30% sludge from recycle paper industry. In order to evaluate the durability during services bulk density, specific gravity, water absorption, mass loss and appearance were determined. Moreover thermal conductivity and compressive were also tested. TGA, XRD and SEM were also carried out to determine the characteristics of raw material. The water abosrbency of the bricks prepared from recycle paper industry sludge and clay was found 10.2%, 13.95% 19.5%, 25.3% respectively, while the density was found 1796 kg/m^3, 1581 kg/m^3, 1444 kg/m^3, 1396 kg/m^3 respectively for 0%, 10%, 20%, and 30 % replacement of clay by sludge. It was observed that bricks prepared from recycle paper industry sludge and clay was light weighted and highly porus possesses sufficient compressive strength and contributes towards solid waste management.

434184 Sustainable Green Composites from Bio-Nylon and Biochar

Abstract:

Green composites derived from crop derived bioplastics and plant derived natural fibres/bio-fillers are the wave of the future. This work deals in the design and engineering of innovative biobased and greener composite materials from the emerging bio-nylons and biomass pyrolyzed co-product, the biochar.

"Use of residues and undervalued co-products from thermo-chemical biomass conversion industries (pyrolysis industries) as the raw materials for new value-added biocomposite material development" is an approach here in enhancing the sustainability of the growing thermo-chemical bioconversion process. Biochar production is a well-known carbon negative or carbon subtractive approach with greener energy solution, which receives more attention due to their environmental impact and other potential uses including soil amendment, metallurgical as well as energy generation. Bio-carbon expects to revolutionize the plastics and related manufacturing sectors with a green alternative through uses of novel biocomposites materials.

Again, renewable resourced-based bionylons (polyamides, PA) have attracted recent attention because these polyamides are based in part with the derivatives of castor oils. Major industry players like BASF, DuPont, Arkema, Evonik and DSM have come-up in bringing a number of bio-nylons (fully or partially derived from renewable resources) like PA 4,10; PA 6,10; PA 10,10; and PA 11 etc. to the commercial/semi-commercial status. Certain specific grades of bionylons have been of great interest to automotive industry due to their better moisture resistance in addition to their superior thermal properties.

The use of biochar as a potential reinforcement in bionylon polymer based biocomposites is investigated in this study. Composites from fractionated miscanthus biochar powder were injection moulded at a constant biochar loading of 20 wt.%. Test samples were characterized for their mechanical (tensile, flexural and impact), heat deflection temperature (HDT) and thermal properties evaluations. Morphological analysis was performed by means of scanning electron microscopy (SEM). Both tensile and flexural moduli exhibited increase resulting from the stiffening effect of the biochar. Flexural strength showed an almost two-fold increase with the addition of biochar in the studied experimental conditions. The surface morphology of the impact fractured surface of the composite revealed good wetting of the biochar with the nylon indicating some level of interfacial compatibility. Compatibility between these materials can be attributed to polar-polar interactions and possibly hydrogen bonding at the molecular level. This presentation will highlight the potential opportunities of such biobased composite materials for certain automotive parts uses.

Acknowledgements: The authors are thankful to the Ontario Ministry of Agriculture, Food and Rural Affairs (OMAFRA)/University of Guelph Research Program under Product Development and Enhancement through Value Chains Research Theme Project # 200399; the Ministry of Economic Development and Innovation (MEDI), Ontario Research Fund - Research Excellence Round 4 Program Project # 050231 & 050289; the Natural Sciences and Engineering Research Council (NSERC), Canada for the Discovery Grants Project # 400322; and the NSERC NCE AUTO21 Project # 460372 for their financial supports.

418949 Chemical Control of Stickies in Recycled Paper Mills

Abstract:

Chemical control of stickies in Recycled Paper mills.

AIChE Annual meeting. Salt Lake City. Nov. 8-13, 2015

Virgilio L. Gonzalez, PhD

Gonzalez Briceño Consultores 75, C.A.

E-mail: gbconsult75@gmail.com

ABSTRACT

Paper manufacture is commonly carried out experiencing some forms of contaminants, affecting process equipment productivity, particularly when Recycled Paper / Card board is used, being stickies among the most critical problems. Many solutions have been proposed, but in our experience, encapsulators can make a big difference in the treatment, compared to common dispersants available in the market.

It is also important to report that there are now Quality Control systems that accurately detect the presence of stickies and the positive effect of over paper quality, when antisticky products, such as encapsulators developed are used. Previous Quality Control systems were not capable to differentiate stickies from ink and other deposits, in Paper making streams.

The differentiation takes place by using a software installed in a conventional PC and a scanner than can measure the number of stickies by surface unit. Features of the QC system are explained and some industrial results are shown , concluding that encapsulators can be a very efficient and economic way to treat stickies and that using the new Quality Control System in Paper mills can validate the product choice made.

425784 Biobased Engineering Plastic Blends for Automotive Application

Abstract:

The problem of introducing the bio-based content in polycarbonate blends was attempted numerous times. Most of these attempts were aimed on poly(carbonate) modification via addition of poly(lactic acid). Typically, these blends have intrinsic poor interphase compatibility resulting in poor impact strength and low heat deflection temperature. Therefore, research was mostly aimed on compatibilization of PC/PLA blends and improving mechanical properties by means of introduction of various additives to overcome brittleness and low heat resistance. In present study PC/PLA blend containing 27-45 wt.% of PLA was modified using specific impact modifier. It was shown that impact modifier concentrated exclusively in brittle PLA phase allowing achieving impact strength exceeding that of pure PC with relatively small amounts of modifier. The addition of PLA to PC allowed to improve overall mechanical properties while maintaining its heat resistance in the ranges over 130 °C

432767 The Influence of Processing Parameters and Use of a Chain Extender on the Mechanical and Thermal Properties of Injection Molded Poly (trimethylene terephthalate) Bioplastic

Abstract:

Poly (trimethylene terephthalate) (PTT), a partly bio-based engineering thermoplastic, was extruded and injection molded and the effect of processing parameters and addition of a chain extender on the engineering properties of the polymer was investigated. The injection molded samples were characterized by thermal, mechanical and microscopic techniques. The heat deflection temperature was found to increase with mold temperature indicating higher and/or more uniform crystallinity of the PTT. In addition, parts molded at higher mold temperatures showed more stable coefficient of thermal expansion below and above the glass transition temperature of the PTT, lending support to the proposed improvement of crystallization by higher mold temperature. The effect of a chain extender additive on the mechanical and thermal properties of the PTT will be presented. The purpose of the study was to optimize the processing conditions for subsequent development of PTT compounds.

Acknowledgement: The financial support from the Ontario Ministry of Agriculture Food and Rural Affairs (OMAFRA)- University of Guelph Product Development and Enhancement through Value Chains Research Theme; the Ministry of Economic Development and Innovation (MEDI), Ontario Research Fund - Research Excellence Round 4 program and the Natural Sciences and Engineering Research Council (NSERC), Canada for the Discovery grants individual (to Mohanty) to carry out this research is gratefully acknowledged.

434937 Synthesis and Characterization of Epoxy Resins Utilizing Plant-Derived Phenolic Acids

Abstract:

Epoxy resins are widely applied in composites, coatings, adhesives, automotive components, wind turbine blades and other applications, due to their superior chemical properties, electrical and heat resistance, adhesion, and mechanical properties. However, most commercial epoxy resins are derived from non-sustainable petroleum, which is a limited resource. For this reason, sustainable biomass derivatives – such as vegetable oils, plant sugars, lignin and cellulose – have attracted much interest to replace traditional epoxy resin components, such as the diglycidyl ether of bisphenol A (DGEBA). The objective of this project is to explore phenolic acids, sustainable, plant-derived molecules, as an alternative to DGEBA. Phenolic acids were chosen as the rigid aromatic rings were expected to contribute to the mechanical strength of the resulting resins, and the presence of carboxylic acid and hydroxyl groups provide ease of functionalization. A diverse array of phenolic acids were epoxidized: salicylic acid, 4-hydroxybenzoic acid, gentisic acid and gallic acid. The number and the relative placement of functional groups within the molecules vary. Relationships were developed between the structure of the phenolic acids, and the resulting network structure, thermal behavior, and mechanical properties of the films.

437464 Study of the Effects of Carbonization Parameters on the Structure of Carbonized Electrospun Lignin

Abstract:

Carbon fibers with various carbon contents, dimensions, and wide range of properties have been developed for multiple applications such as light-weight advanced composite reinforcements for aerospace applications, automotive parts, oil exploration tools, or for alternate energy harvesting, storage, or conversion, electrodes, sensors, etc. Conventional carbon fibers are produced from polyacrylonitrile (PAN) or petroleum-based pitch. Production of high performance carbon fibers from PAN is a high cost process, which limits the application of this material in its growing market. The growing carbon fiber market demands sustainable products at lower costs with various functionalities to meet requirements of different applications.

Raw material and processing costs are the two main elements of the high cost of PAN-based carbon fibers. Alternate raw materials and processing techniques have been studied for producing wide range of products which could be optimized for certain applications.

Lignin is a natural polymer available in the plants and is a coproduct of paper and cellulosic ethanol industries. It has been widely studied as a renewable low-cost precursor for producing carbon materials. Several spinning techniques have been studied for production of fibers from lignin such as dry-spinning, melt-spinning, and electrospinning.

Electrospinning of lignin is a versatile process which can be used to spin solutions of lignin, lignin – polymer blends, or lignin with particle fillers. Electrospun fibers can be collected as randomly oriented or aligned fiber mats with fiber diameters in the range of sub-micron to a few micro-meters.

In this study, electrospun fiber mats of lignin were produced by spinning an alkaline aqueous solution of lignin and poly(ethylene oxide) with the ratio of 95/5 wt%. For comparison of the properties, electrospun PAN fibers were also produced by spinning of solutions of PAN in N,N-dimethylformamide. Carbonization of the fibers was done in a two-step thermal treatment. The first step, thermostabilization, was done by heating the fibers in a tube furnace in presence of air from room temperature to 250 °C and kept at that temperature for two hours. The second step, carbonization, was the main subject of this research. The lignin and PAN fibers were carbonized at different combinations of conditions. The carbonization conditions were selected by applying factorial design of experiments. The studied variables were heating rate, carbonization temperature, and time. Morphology of the carbonized fibers was studied by scanning electron microscopy. Lignin fibers spun from aqueous solutions had about 25% reduction in their diameter after carbonization. In contrast, PAN fibers melted and created a network of fibers at high temperatures (1100 °C) and longer times of 10h. The characteristics of carbon structure of the fibers were studied by Raman spectroscopy. It showed that the R ratio, which is a measure of degree of ordered graphitic structure, for PAN is strongly dependent on the temperature. In comparison, for lignin fibers, the interaction between temperature and heating rate showed the highest influence. The same trend happened for graphitic crystalline size. Carbon structure of the fibers has effect on the final properties. Therefore, the knowledge obtained on the effects of carbonization conditions on the carbon structure is useful to change the structure and optimize the properties for specific applications.

Acknowledgement

The authors are thankful for the financial support from the Natural Sciences and Engineering Research Council (NSERC), Canada, for the Discovery grants individual (to Misra), NSERC NCE AUTO21, the Ontario Ministry of Economic Development and Innovation (MEDI) for the Ontario Research Fund (ORF) Research Excellence (RE) Round-4, and the Ontario Ministry of Agriculture, Food and Rural Affairs (OMAFRA) - University of Guelph Bioeconomy Industrial uses research program. The authors are also thankful to Lignol Innovations Ltd. for providing the Organosolv Lignin samples

(332a) Using a Combined Hydrolysis Factor to Optimize High Titer Ethanol Production from Sulfite-Pretreated Poplar without Detoxification

Zhang, J., Beijing University of Chemical Technology
Zhu, J. Y., USDA Forest Service, Forest Products Laboratory
Gu, F.,
Zalesny, Jr, R.,
Originally presented on: 11/10/2015 12:30:00 - 12:55:00

Sulfite pretreatment to overcome the recalcitrance of lignocelluloses (SPORL) was applied to poplar NE222 chips in a range of chemical loadings, temperatures, and times. The combined hydrolysis factor (*CHF*) as a pretreatment severity accurately predicted xylan dissolution by SPORL. Good correlations between *CHF* and pretreated solids enzymatic digestibility, sugar yield, and the formations of furfural and acetic acid were obtained. Therefore, *CHF* was used to balance sugar yield with the formation of fermentation inhibitors for high titer ethanol production without detoxification. The results indicated that optimal sugar yield can be achieved at *CHF* = 3.1, however, fermentation using un-detoxified whole slurries of NE222 pretreated at different severities by SPORL indicated *CHF* » 2 produced best results. An ethanol titer of 41 g/L was achieved at total solids of approximately 20 wt% without detoxification with a low cellulase loading of 10 FPU/g glucan (19 mL/kg untreated wood).

435439 Advanced Biological and Chemical Design for Lignin Bioconversion

Abstract:

The utilization of lignin for fungible fuels and chemicals represents one of the most imminent challenges in modern biorefinery. However, bioconversion of lignin is highly challenging due to its recalcitrant nature as a phenolic heteropolymer. We first addressed the challenges by revealing the chemical and biological mechanisms for synergistic lignin degradation by a bacterial and enzymatic system, which significantly improved lignin consumption, cell growth and lipid yield. The *R. opacus* cell growth increased exponentially in response to the level of laccase treatment, indicating the synergy between laccase and cells in lignin degradation. Other treatments like iron and hydrogen peroxide showed limited impact on cell growth. Chemical analysis of lignin under various treatments further confirmed the synergy between laccase and cells at chemical level. ^{31}P NMR suggested that laccase, *R. opacus* cell and Fenton reaction reagents promoted the degradation of different types of lignin functional groups, elucidating the chemical basis for the synergistic effects. ^{31}P NMR further revealed that laccase treatment had the most significant impact for degrading the abundant chemical groups. The results were further confirmed by the molecular weight analysis, the HSQC analysis and lignin quantification by Prussian Blue assay. The cell-laccase fermentation led to a 17-fold increase of lipid production. Overall, the study indicated that laccase and *R. opacus* can synergize to degrade lignin efficiently, likely through rapid utilization of monomers generated by laccase to promote reaction toward depolymerization. Based on the study, consolidated lignin conversion was engineered using laccase and *R. opacus*. An efficient ligninase secretion system has been established through proteomics-guided optimization of promoter/rbs, secretion peptide, and others. The advanced biodesign has achieved over 70U/mL of laccase production and nearly 1000 fold increase of cell growth. The integration of biological and chemical design has led to the lipid titer of over 1g/L on insoluble Kraft lignin. The study has paved a path for more efficient conversion of lignin and waste material from biorefinery.

427551 Physio-Chemical Characterization of 3D Structure of Plant CELL Walls

Abstract:

Plant biomass is a renewable, sustainable raw material for a wide variety of bio-based based products including ethanol, paper, polylactic acid, wood plastic composites etc. Plant cell walls have a complex internal porous structure that has a significant influence on the efficiency of the biomass conversion process, the properties of the materials and their end-use applications. When subjected to pretreatment, one or more of the cell wall components degrade altering the internal structure and thus affect further conversion processes. With the advent of novel materials characterization techniques it is now possible to non-intrusively visualize the internal three dimensional structure of porous materials and attempt to better understand structure property functional relationships. Techniques such as Raman Spectroscopy, Atomic Force Microscopy, and Scanning Electron Microscopy etc. provide topochemical and structural information. But in order to get a true sense of the internal ultrastructure of the cellwall 3D visualization techniques are required. Previous work on characterization of 3D porous biomaterials was accomplished using X-ray computed tomography which provides excellent visualization at the micron scale. Even though micron scale structural details may be sufficient in many applications of biomaterials, in order to better understand the ultrastructure of plant cell walls and their role in biomass conversion processes, it is necessary to probe the ultra-structural features at the nanoscale. The 3D computed tomography technique mentioned above is now being extended to other imaging techniques such as Transmission Electron Microscopy (TEM). It provides an X-Y resolution of upto 1-2nm. The technique is based on capturing a series of images by successive tilts around a single axis and reconstructing the 3D image by back projection.

Using the information from TEM computed nanotomography (TEM-n-CT), the porosity, pore size distribution, interfacial area, structural tortuosity and diffusion characteristics of the biomass can be determined. The cell wall material can be considered to be made up of two phases- namely void spaces comprising of cell lumen, pits etc. and a solid cell wall comprising of cellulose microfibrils embedded in lignocellulose polymer matrix It is also possible to track the changes in the 3D structure asthe biomass undergoes pre-treatment and bioconversion processes. . The effect of pre-treatment processes using hot water, acid, and alkali etc on the 3D ultrastructure of biomass can be explored.

In addition, using the actual 3D structure of biomass we propose a simultaneous transport and reaction engineering modeling using stochastic dynamic approach and survival probabilities. This will include random walk simulations in cell wall structures to simulate the diffusive transport of reactants in the pores with possible adsorption and reaction at the cell wall surface following thermodynamic and reaction kinetics considerations of individual biomass components. This will help gain fundamental insights on biomass recalcitrance and how they can be addressed in developing effective conversion strategies.

437221 Phosphorus Recovery from Liquid Animal Manure by Electrocoagulation

Abstract:

Dairy manure, containing high levels of volatile solids, N, P and K that are necessary nutrients for plant growth, is conventionally used as orgnic fertilizer. Removing liquid from the manure solids, by natural sedimentation or mechanical separation, can enrich volatile solids and nutrient content, and therefore decreases its transportation cost for land applications. However, the separated liquid phase still contains soluble and colloidal forms of phosphorus and direct discharge of the manure liquid may cause eutrophication. Collecting the remaining phosphorus from the liquid will increase the fertilizer value of dairy manure and eliminate the issues associated with nutrient contamination. Proper methods should be developed to recover the remaining phosphorus which is difficult to be physically separated from the manure liquids. Electrocoagulation (EC) is a method suitable to decrease pollutants that are difficult to be removed by filtration or centrifugation. So in this work, the possibility of enriching P from the liquid manure was evaluated. First, four electrode materials, including 304 stainless steel, aluminum, low carbon steel and grey cast iron, were compared for their effects in P removal from liquid dairy manure (filtrate of 295-μm mesh screen). Before electrocoagulation, about 80% of solids (on dry basis) in the liquid dairy manure were smaller than 45 μm. After EC treatment, manure particles of small sizes were coagulated and a natural sedimentation for 20 min successfully separated most of the P-bearing solids, evidenced by the increased proportion of particles larger than 250 μm. Although aluminum showed the best P removal performance, aluminum anode leaded to the formation of a thick oxidation film, which stopped the further release of coagulants. Among the iron-based electrodes, the best P removal efficiency of 90.2% was achieved in 100 min by stainless. However, its iron content is only 53.5-74.5% and the other components like chromium and nickel will bring further environmental pollution. Low carbon steel, removing 87.1% total phosphorus within 100 min, had a better P removal efficiency than cast iron. This material also had high iron content and low electrical resistivity, and therefore was used as the electrode material for further exploration. Second, experiments were conducted to optimize the EC process for P removal and to monitor several relevant property changes of sludge and supernatant, such as pH, solid profile and content, iron content, et al. The operating parameters included current (0.2-1.4 A), initial pH (5-9), aeration (0-120 min treatment), agitation speed (0-300 rpm) and anaerobic digestion (manure was digested for different duration before experiments). It was found that alkaline adjustment of the liquid manure significantly improved phosphorus removal. Higher current produced coagulants at a higher rate at anode but consumed more electrical power, and thus a cost analysis was carried out to clarify the relationship between the energy consumption and the P removal in various operating conditions. Aeration treatment increased pH and stripped out carbon dioxide and hydrogen sulfide from liquid manure, which was found to have accelerated the P removal in the following EC treatment. Aeration for 60 min at a rate of 0.8 L/min before electrocoagulation process had the best P removal performance. Agitation could increase the mass transfer of coagulants in the medium but meanwhile the coagulated particles would be broken down when agitation speed was too high. As a result, the agitation rate of 75 rpm was found to have the best P removal rate. Anaerobic digestion (AD) could reduce the organic matter in manure and thus reduce the solid content. It is assumed that the energy provided during EC process has an effect on both P removal and solid coagulation. More energy needs to be contributed to P removal due to the low solid content in anaerobic digested manure. 97.26% of total phosphorus was removed within 70 min when using the manure after anaerobic digestion for 10 days. Finally, empirical models were established to predict the treatment performance, reaction time and energy consumption at given operating conditions. The EC method developed in this study successfully achieved the goal of expediting manure solid/liquid separation and P recovery. The produced EC sludge will be evaluated for fertilizer use.

427462 Estimation of Energy Consumption and Production in Woody Biomass Pulverization with a Multiple Tube Vibration Mill

Abstract:

Scale-up estimation of woody biomass pulverization with newly developed multiple tube vibration mill (MTVM) was carried out. Pulverization ability in large size MTVM was estimated based on a lab.-scale pulverization experiment. In this estimation, pulverization pot arrangement was changed for the scale up, and optimum pulverization condition of woody biomass in MTVM was calculated in terms of productivity of woody powder and consumed energy per production. To understand the MTVM capability, pulverization performance of MTVM was compared with that of a conventional single pot vibration mill. MTVM have a big advantage over a conventional vibration mill in woody biomass pulverization, and the difference of the MTVM pulverization capability become more significant with increasing the pot size. Applying stage graining method, low crystallinity woody powder (crystallinity index <10%) can be produced around 1.2 t/h with 1.0 GJ/t by an industrial-scale MTVM. Consequently, total energy consumption in woody biomass pulverization process including coarse grinding, middle grinding and fine grinding, became 1.48 GJ/t.

437349 Lime Sludge: An Emerging Alternate Construction Building Material for the Partial Replacement of Fine Aggregate

Abstract:

Lime sludge genrated in chemical recovery section of a paper mill is recoverd through brurning lime kiln. Lime sludge generated from agro based paper mill poses problem in lime kiln due to high amount of silica. There is need to seek its utilisation in various other industrial applications as secondary raw materials. This present study is focused on the utlisation of lime sludge in making concrete blocks. The experimental investigations are concerned with strength of concrete and optimum percentage of the partial replacement by replacing fine aggregate via 25 %, 30 %, 35%, 40 % and 50% of lime sludge. The mixture of concrete of lime sludge, fine aggregate, coarse aggregate, ordinary portland cement, polymeric binders, and water were prepared by using standard consistency method as per IS: 10262-2009. The dimensions of cubes were 100mm X 100mm X 100mm. Different concrete mixes by using lime sludge, fine aggregate, coarse aggregate, and cement in the varying proportions were prepared with water to binder w/(c + LS) ratio of 0.48 to 0.80 for target strength of 37 (N/mm^2). The compressive strength of the cubes of concrete is reduced with the replacement of lime sludge of pulp and paper mill over the control cement cube concrete. However with the use of plasticisers the strength of concrete bloacks can be enhanced to a stisfactory level with the lime sludge replacment upto 30%. The effect of compressive strength at 7 days and 28 days has been discussed here.

417382 A New Method to Extract High Purity Lignin from Woody Biomass

Abstract:

In this presentation we will report recent research results on applying Deep Eutectic Solvents (DES) to extract lignin from woody biomass in high yields and high purity. The treatments were conducted between 90°C and 180°C. The lignin extraction yield up to 78% was obtained with purity about 90% based on Klason lignin content. Strategies to optimize lignin extraction and DES recovery will be discussed in this presentation.

423230 Technical Feasibility Evaluation of Alcohol-to-Jet (ATJ) Conversion Technology for Renewable Jet Fuel

Abstract:

Abstract Submission to Be Considered As An Oral Presentation at AIChE 2015 Annual Meeting in the Biochemical Conversion Processes in Forest/Plant Biomass Biorefineries Session

Technical Feasibility Evaluation of Alcohol-to-Jet (ATJ) Conversion Technology for Renewable Jet Fuel

Scott Geleynse, Manuel Garcia-Perez, Michael Wolcott, Xiao Zhang

Washington State University

The FAA is working to enable the U.S. use of one billion gallons per year of "drop-in" sustainable alternative jet fuels by 2018. Airlines and aerospace manufacturers around the world have also expressed continuously increasing interest in the replacement of jet fuel from fossil fuels to renewable fuels from biomass. Despite the considerable investment in the research of several different conversion strategies, establishing the successful commercial production of biorenewable jet fuel remains a challenge. In this presentation, we will review strategies to use alcohols as a platform molecule to generate jet fuel and other products. By highlighting the technical opportunities and barriers in the process, we aim to provide insight into the various political and economic challenges faced in making and meeting renewable fuel goals.

Presenting author:

Scott Geleynse

Bioproducts Sciences and Engineering Laboratory

School of Chemical Engineering and Bioengineering, Washington State University
2710 Crimson Way, Richland, WA 99354

Contact: scott.geleynse@email.wsu.edu, x.zhang@wsu.edu

428042 Dilute Alkali Treatment Followed By Mechanical Refining to Enhance Sugar Conversion

Abstract:

Lignocellulosic biomass has acetate groups in xylan backbone, which can be released during pretreatment process and negatively affect the microbial growth during the fermentation process. It has been reported that the deacetylation of biomass with alkaline reagents can remove both acetate groups and partial fraction of lignin, and enhance the carbohydrate conversion yield in a subsequent enzymatic hydrolysis, as well as the efficiency in the downstream fermentation process. Meanwhile, mechanical treatment is recently considered as a promising technology to enhance the yield of carbohydrate conversion in pretreated lignocellulosic biomass. The purpose of this work is to optimize deacetylation process for corn stover with minimum loading of NaOH with a post-treatment of mechanical refining. Due to cost burden of NaOH and its recycling process, the decrease of NaOH loading is critical for economic feasibility of the whole conversion process. Preliminary data presented that deacetylation at 80 °C for 20 min with 2% NaOH (w/w of biomass) could remove 70.6% of acetate groups in raw corn stover, and at the elevated temperature of 140 °C for the same reaction time and NaOH loading, 74.0% of acetate groups were removed. With less than 4% difference in changes of acetyl content, the resulting in carbohydrate conversion yield showed substantial changes between the two deacetylated biomass, 35.8 and 52.2%, respectively. When they are subjected to mechanical treatment, the conversion yields for both samples were improved to 50.0 and 65.0%, respectively. To mitigate the decline in pretreatment yield from lower NaOH loading, the effect of pretreatment at the elevated temperature and mechanical refining under higher temperature/pressure will be also considered in this study.

429399 Investigation of Dissolved Lignin in Autohydrolysis Liquor

Abstract:

Autohydrolysis (water, temperature, pressure) pretreatment of lignocellulosic biomass is a sustainable unit operation that effectively reduces biomass recalcitrance for downstream sugar production in a biorefinery. The liquor stream produced is enriched with dissolved hemicellulose and lignin compounds with a distribution of molecular weights. Valorization of this stream will increase the economic viability of the lignocellulosic biorefinery. While dissolved hemicellulosic sugars are well-characterized in terms of carbohydrate structure and degree of polymerization, dissolved lignin has received less attention mainly due to the challenge in characterization. The dissolved lignin is of particular interest since up to 35% of non-wood lignin and 19% of hardwood lignin can be made water soluble during autohydrolysis. Understanding the dissolved lignin will lend more knowledge toward valorization pathways for the autohydrolysis liquor. Extractive-free non-wood sugarcane bagasse and hardwood sweetgum (*L. styraciflua)* biomasses were subjected to identical autohydrolysis pretreatments at 180 °C for 40 minutes and the liquor stream was collected through vacuum filtration. Quantitative GC-MS analysis of the monomeric phenolics in autohydrolysis liquor showed that only ~1% of the dissolved lignin is monomeric, leaving the remaining dissolved lignin uncharacterized. The predominant lignin monomers in non-wood liquor are *para*-coumaric acid and vanillin with a total monomer concentration of ~150 mg/L, while in hardwood liquors the dominant molecules were syringaldehyde and vanillin and a total monomer concentration of ~45 mg/L. The liquor was then dried to produce dissolved solids residue followed by solvent extraction to collect the lignin. The experimental results from high-resolution MS for chemical structure, GPC for molecular weight, and NMR for inter-lignin bonding structures will be discussed.

433029 Process Modeling of a Distributed Sugar Depot Approach to Biomass Conversion

Abstract:

The development of commercially viable, integrated biorefineries remains a difficult challenge. Converting biomass through distributed "sugar depots" offers a flexible alternative route, with advantages in local feedstock access and the use of mechanical biomass deconstruction only available at smaller facilities.

In this study, we will present a sugar depot process model detailing the milling and hydrolysis of Douglas fir residuals to produce liquid sugars and byproducts. We evaluate and compare several process scenarios via technoeconomic analysis and identify opportunities for process cost reduction.

435445 Simultaneous Conversion of All Cell Wall Components By Oleaginous Fungus without Chemi-Physical Pretreatment

Abstract:

Lignin utilization during biomass conversion has been a major challenge for lignocellulosic biofuel. In particular, the conversion of lignin along with carbohydrate for fungible fuels and chemicals will both improve the overall carbon efficiency and reduce the need for chemical pretreatments. However, few biomass-converting microorganisms have the capacity to degrade all cell wall components including lignin, cellulose, and hemicellulose. We hereby evaluated a unique oleaginous fungus strain *Cunninghamella echinulata* FR3 for its capacity to degrade lignin during biomass conversion to lipid, and the potential to carry out consolidated fermentation without chemical pretreatment, especially when combined with sorghum (*Sorghum bicolor*) *bmr* mutants with reduced lignin content. The study clearly showed that lignin was consumed together with carbohydrate during biomass conversion for all sorghum samples, which indicates this organism has the potential for biomass conversion without chemical pretreatment. Even though dilute acid pretreatment of biomass resulted in more weight loss during fungal fermentation than untreated biomass, the lipid yields were comparable for untreated *bmr6/bmr12* double mutant and dilute acid-pretreated wild-type biomass samples. The mechanisms for lignin degradation in oleaginous fungi were further elucidated through transcriptomics and chemical analysis. The studies showed that in *C. echinulata* FR3, Fenton Reaction may play an important role in lignin degradation. This discovery is among the first to show that a mechanism for lignin degradation similar to ones found in white and brown rot basidiomycetous fungi exists in an oleaginous fungus. This study suggests that oleaginous fungus such as *C. echinulata* FR3 can be employed for complete biomass utilization in a consolidated platform without chemical pretreatment, or can be used to convert lignin waste into lipids.

Mixing Effects on the Reaction Kinetics and Dynamics of *Saccharomyces Cerevisiae* Mediated Glucose Fermentation to Bioethanol

Ashwin Gaikwad, Department of Chemical Engineering, Visvesvaraya National Institute of Technology, Nagpur-440010, India

Saikat Chakraborty, Department of Chemical Engineering, Indian Institute of Technology Kharagpur, Kharagpur- 721302, India

Abstract

Conversion of lignocellulosic biomaterials to valuable fuels and chemicals is seen to be highly attractive because of the abundantly available lignocellulosic biomass. To convert these biomaterials into fuels, enzymatic hydrolysis and fermentation are proved to be most promising and environmentally benign processes. This work explores the effects of mixing on the reaction kinetics and dynamics of *Saccharomyces cerevisiae*-mediated fermentation of glucose to bioethanol at four different mixing conditions, i.e., 0 rpm (no mixing), 40, 80 and 150 rpm (high mixing). Based on the idea that mixing is one of the important process parameters to influence bioethanol yield in fermentation, we explore the effect of mixing on the reaction kinetics and dynamics of fermentation process using a coupled experimental and modeling approach. The growth kinetics of *S. cerevisiae* was shown to be influenced by the mass transfer mechanism and μ_{Max} and K_s were both increases consistently with increase in the mixing speed. As the domination of ethanol formation was seen only for the first 12 hours and after that ethanol consumption by *S. cerevisiae*, attributes to the metabolic and biomass structured model of glucose metabolism via catabolic and anabolic pathways. The kinetic constant for ethanol formation (k_1) increases with the mixing speed and attains a constant value at 80 rpm whereas the kinetic constant for ethanol consumption (k_2) increases monotonically up to 100 rpm and attains constant value thereafter. At a mixing speed of 80 rpm, the reactor operates in the intermediate regime, where both mass-transfer and reaction kinetics govern the product yield, and we obtain the maximum ethanol yield (4.32 mg/ml), which is 89% of the theoretical value. Thus, it was discovered that the maximum ethanol yield was achieved in neither of the two asymptotic limits of mass transfer limited asymptote (no mixing) or reaction limited asymptote (complete mixing), but at moderate mixing conditions (in the intermediate regime where both mass-transfer and reaction play determining roles). A mathematical model developed for the proposed reaction scheme for the determination of ethanol yield at different mixing conditions matches well with the experimental data confirms the validity of our model for the batch reactor configuration.

Keywords: Lignocellulose, Mixing, Fermentation, Glucose, Mass Transfer, Ethanol

421286 Study of Structural Characteristics of Lignin Fractions Obtained from Sequential Solvent Fractionation

Abstract:

Lignin, one of abundant polymeric materials in Earth, has heterogeneous and complex chemical structure, which can be big barrier for its valorization. Thus, recent researches of lignin valorization are interested in improving its homogeneity, and sequential lignin fractionation with organic solvents can be considered as one of the solution to overcome this problem. Because of different dissolving capacity of organic solvents, lignin can be separated as well as homogenized according to solvent fractionations. Therefore, the purpose of our study was to fractionate various lignins by molecular weight and analyze structural characteristics of them for their efficient utilization. In our study, milled wood lignin(MWL), soda lignin (SL), and organosolv lignin (OL) was fractionated successively with ethyl acetate (F1), methanol (F2), acetone (F3), and dioxane/water (95:5v/v, F4), and insoluble fraction was named for F5. The yields of each fraction of MWL were 7.7%, 26.7%, 9.2%, 56.4% with the absence of F5, whereas those of SL were 22.0%, 52.5%, 4.6%, 14.9%, 6.1%. In the case of OL, only F1, F2, and F4 were obtained 20.8%, 62.1%, and 17.1%. In GPC analysis, as dissolving capacity of solvent increased, its soluble fraction showed higher average molecular weight (M_w) in all lignins, only F5 of SL was not able to be analyzed due to its insolubility to GPC solvent. Methoxyl group contents of lignin fractions tended to decrease as compared with those of raw lignins, but didn't show any comparable tendency between them. Thermal degradation properties, determined by thermogravimetric analysis, showed that thermal stability of MWL fractions showed no specific differences between them, while that of SL fraction decreased as molecular weight increased. In the case of OL, thermal stability of F4 seemed to significantly decrease compared to other fractions. For detailed study of their structures, each fraction needs to be also introduced to other analyses such as DFRC, Pyrolysis-GC/MS, FT-IR and 2D-HSQC-NMR.

427642 Green Coal Production from Biomass

Abstract:

Green Coal Production from Biomass

Stanislav Barskov[1], Dr. Prashanth R. Buchireddy[2], Dr. John L. Guillory[2], Dr. Mark E. Zappi[1]

[1] Department of Chemical Engineering University of Louisiana Lafayette, Lafayette, LA

[2]Department of Mechanical Engineering University of Louisiana Lafayette, Lafayette, LA

Coal, one of the most abundant natural resources has been extensively utilized for U.S. power generation in the last fifty years. Today, over 50% of U.S. power is generated by combustion of coal in the power plants. When combusted in the power plants, coal emits substantial amounts of greenhouse gases into the atmosphere. As a result, federal government has mandated environmental agencies to establish strict regulations and propose solutions on the reduction of the greenhouse gases emissions. One of the emerging solutions to reduce greenhouse emission is co-firing biomass with coal. The main advantages of the biomass is that it is considered carbon neutral and has a lower nitrogen, sulfur, and ash content than coal. The lower content of nitrogen, sulfur, and ash results in significantly lesser amount of emissions when biomass is combusted. The disadvantages of co-firing biomass are that biomass has lower heating value, density, and higher moisture content when compared to coal. Additionally, difficulties with fuel preparation, handling, storage, transportation, milling, and feeding in for co-firing with coal in the power plants further hinder biomass utilization as an alternative energy source.

One of the way to address the aforementioned issues associated with biomass co-firing is through torrefaction. Torrefaction is a thermal pre-treatment of the biomass in temperatures between 230-320 °C in an absence of an oxidizing agent and under atmospheric pressure. During torrefaction approximately 30% of the biomass is converted into volatile organic compounds and gases containing 10% of biomass energy. However, 70% of the mass and 90% of the energy is still retained in the original biomass. Therefore, energy densification is achieved via torrefactions by a factor of 1.3 on mass basis. The torrefied product has improved fuel properties such as increased mass energy density, hydrophobicity, uniformity, and grindability compared to the fresh biomass. This properties allow for effective utilization of torrefied biomass in coal fired power plants using existing burners and without major modifications.

Research is being conducted at University of Louisiana Lafayette using an indirectly heated rotary drum pilot scale torrefaction reactor utilizing pine wood chips as a primary lignocellulosic material. The effect of various parameters including temperature, raging in between 240 - 320 °C, residence time, ranging in between 30 - 100 minutes, carrier gas (N_2, CO_2, Steam), particle size, moisture content, ranging in between 10 - 30% on torrefaction product is being evaluated and the results will be presented in this conference. Additionally, analysis of volatiles and gases is being carried out to evaluate the overall process efficiency. Lastly, mass and energy balances on the pilot scale torrefaction process would be presented. Conducting research on a pilot scale and utilizing alternative energy sources would put U.S. one step closer to achieving perfection in clean power generation, and establishing energy independence from the rest of the world.

433618 CO2 Enhanced Gasification of Biomass Char - Catalytic Effects of Na, K, Ca, and Mg

Abstract:

Biomass can be converted to biofuels using various pathways which are broadly classified as biochemical or thermochemical. Gasification is a common thermochemical process that converts biomass to a gaseous product, syngas, which can be used for power generation, chemical production and as a fuel. The use of air, steam and oxygen as oxidizing agents in this process has been widely studied and reported. Catalytic gasification of biomass can effectively improve the reaction rate and conversion efficiency compared to conventional gasification. However, very little is known about the catalytic gasification of biomass char with CO_2.

In the current study, unloaded biomass char was gasified using CO_2 at five different temperatures between 800°C – 950°C in a fixed bed reactor. Using the evolution of CO in the gasification process as basis for conversion, the conversion-time data was fitted to three different solid-gas reaction models. The Random Pore Model (RPM) gave the best fit and the kinetic parameters for the same were calculated. In the second part of the study, acetates of Sodium (Na), Magnesium (Mg), Potassium (K) and Calcium (Ca) are used to load the char with metal catalysts and subjected to the same gasification procedure. The effect of catalysts is studied using the variation in kinetic parameters obtained using the RPM.

Instruments such as a CHNS analyzer, inductively coupled plasma (ICP), Scanning Electron Microscope (SEM) are used to analyze the composition and properties of char. Additionally, a micro GC is used to analyze the syngas at the output.

434043 Visualization of Miscanthus x Giganteus Cell Wall Deconstruction Subjected to Dilute Acid Pretreatment for Enhanced Enzymatic Digestibility

Abstract:

Natural recalcitrance that arises from the complicated cell wall structure and heterogeneous components distribution impedes the deconstruction of lignocellulosic plant cell walls to fermentable sugars. Dilute acid pretreatment (DAP) is an attractive method to overcome the ultrastructural and chemical barriers for rendering enzymatic conversion of polysaccharides. In this study, *Miscanthus* x *giganteus*internodes serving as a model bioenergy crop was subjected to DAP with increasing severity to obtain a range of samples with altered structure and chemistry.

DAP of *M.* x *giganteus* resulted in solubilization of arabinoxylan and cross-linking hydroxycinnamic acid in a temperature dependent manner. The optimized pretreatment at 170 °C, 1% $H_2SO_4$for 30 min enabled a significant enhancement in sugar release efficiency of the treated samples in 72 h compared to untreated samples (51.2% yield compared to 11.8%, an increase of 4.4-fold). This remarkable improvement was favored by a sequence of changes that occur in plant cell walls during pretreatment-induced deconstruction, namely, the loss of matrix between neighboring cell walls, selective removal of hemicelluloses, redistribution of phenolic polymers, and increased exposure of cellulose. In addition, changes in cell wall structure including opening up the cell wall, increase of porosity and decreasing mechanical resistance were also found to enhance enzyme access to cellulose and further sugar yield.

Dilute Acid Pretreatment is an effective approach that enhancesthe bioconversion of cellulose to glucose by breaking down the macroscopic rigidity and microscopic resistance of cell walls.

431771 Catalytic Fast Pyrolysis of Lignin Recovered from Hot-Water Extract of Electron-Beam-Irradiated Sugar Maple for the Production of Low Molecular Weight Phenolic Compounds

Abstract:

Thermochemical degradation of lignin at a high temperature in a controlled atmosphere (pyrolysis) is a promising technique to replace fossil fuels for the production of low molecular weight (LMW) phenolic chemicals. Conventionally, however, pyrolysis of lignocellulosic (LC) materials including current sulfur-containing technical lignins (e.g. kraft, lignosulfonates) is plagued by low yields, poor selectivity to desired products, and catalyst deactivation. These issues must be addressed for lignin pyrolysis to become a viable industrial process. It has been shown that the nature of the material being pyrolyzed has a large impact on the resulting yield of phenolic products. Furthermore, it has been shown that biorefinery lignins may have a more favorable structure and composition (e.g. increased purity and lower polydispersity) for the production of high-value products as compared to kraft lignin. In this work, results of pyrolysis and catalytic upgrading of novel biorefinery lignins are presented to better understand the effects of lignin structure and composition on pyrolysis products. Electron beam irradiation (EBI) has been shown to increase mass removal during autohydrolysis-based pre-extraction (hot-water extraction). Lignin recovered from hot-water extraction of native and EBI (100-1000 KGy) sugar maple (precipitation at pH 2 followed by centrifugation) was characterized (purity, free phenolic hydroxyl group content, molecular weight distribution, S/G ratio, β-O-4 content). Thermal degradation characteristics of these lignins were analyzed by TGA. Fast pyrolysis (550°C) was performed with subsequent catalytic defunctionalization (upgrading) with various zeolite catalysts (ZSM-5, Y, and Beta) to produce a more uniform stream of LMW phenolic compounds; GC-detectable pyrolysis products were analyzed by GC-MS. Based on results, the effect of EBI on lignin structure and properties and the repercussions for catalytic fast pyrolysis will be assessed.

435328 Alcohol Stabilization of Bio-Oils during High Temperature Treatment

Abstract:

Bio-oil produced by fast pyrolysis of biomass requires significant modification to become an acceptable transportation fuel. The presence of large amounts of oxygenated compounds, such as water, carboxylic acid, aldehyde and phenol, is the primary reason for the differences in physical and chemical properties between fossil fuels and bio-oils, which renders bio-oil incompatible with existing fossil fuel refinery processes. Two effective routes in upgrading bio-oils through chemical pathways (deoxygenation) are hydrotreating (~250-450 °C) and catalytic cracking (~450 °C). The processing of bio-oils in existing hydrotreatment and fluid catalytic cracking (FCC) facilities is appealing; however, the high acid content of crude bio-oil may lead to infrastructure corrosion. Additionally, severe aging reactions occur when bio-oil is heated to above 200 °C. The low hydrocarbon yield, formation of undesirable products (e.g., coke, tar, char), catalyst deactivation, and reactor plugging are significant barriers to bio-oil commercialization. To address these issues, conversion of the carboxylic acids and reactive carbonyl compounds into esters, ethers and acetals via the addition of low-cost alcohols is promising as a method to pretreat bio-oil before hydrotreating and catalytic cracking. In this study, methanol, ethanol, 1-propanol, 2-propanol and 1-octanol were added to a crude bio-oil and the mixture were stored at 200 °C for different periods (6-50 h) to investigate changes in bio-oil's physicochemical properties. Treated bio-oils were characterized to by Karl Fisher titration, viscometer, FTIR, GPC, TGA, DSC and GC/MS. Alcohol additives instantly reduced the viscosity of bio-oil and significantly decreased the rate of aging. Methanol showed a unique stabilization effect—a ~10 wt% increase in high molecular mass lignin content was observed for methanol/bio-oil mixtures whereas only slight increases were observed for other alcohol treatments. A rapid decrease in FTIR peak intensity for unconjugated carbonyls for the methanol/bio-oil mixture indicated methanol was more reactive than other alcohols. The C/O ratio was significantly improved by 1-octanol addition, but it resulted in the lowest stabilization. Chemical reactions between bio-oil and the additive alcohols, as well as additional hydrogen-bonding, appears to inhibit or mitigate the aging reactions.

437516 Furnish Optimization for a Multilayered Paperboard

Abstract:

Multi-layered symmetrical paperboard sheets were made in a laboratory sheet former from old newsprint (ONP) in the middle layer and old corrugated container (OCC), Indian mixed recycled paper IMR), and virgin bleached softwood kraft (BSK) pulp in surface layers. The proportions of ONP and surface layer pulp were varied. Cationic starch in different proportions was added to the pulps used. The board sheets were surface sized in a laboratory coating unit with a mixture of oxidized starch and sodium silicate in varying proportions.

Different strength properties such as tensile strength, burst strength, tear strength, bending stiffness, compression strength (ring crush test), and ply-bond strength (Scott Bond test) of these board sheets were determined using standard testing procedures. In the pulps used, the BSK had much higher tensile, burst, and tear strength than those of OCC, IMR, and ONP. The ONP was the bulkiest of all the pulps and produced the sheets with maximum bending stiffness index. Tensile strength of OCC, IMR, and ONP were nearly equal, the OCC and IMR had slightly higher tear strength than the ONP. The major observations were:

- Tensile, burst, tear, and compression strength of the board varied almost linearly with the proportions of the pulps used irrespective of the construction of the board, whether multilayered or single-layered form the mixture of the pulp.
- Bending stiffness passed through a maximum for symmetrical multilayered board when about 70% ONP was used in the middle layer and 30% of either, BSK, OCC, or IMR was used in the surface layers. The improvement achieved in bending stiffness was nearly identical for either of BSK, OCC, and OMR in the surface layers.
- The Scott Bond values were also depended on the proportions of the pulps used in the board in a manner similar to tensile and tear strength. However, for a given proportions of pulps, Scott Bond values were slightly higher for multilayered board than for the single-layered board.
- Addition of cationic starch to the stock resulted in a significant improvement in Scott Bond values of multilayered board.
- Surface sizing of the board with oxidized starch resulted in a significant improvement of RCT values of the board. The increased proportion of sodium silicate in the sizing solution further improved the compression strength.

Relationships between the physical and morphological characteristics of the component pulps and the board properties have been established to enable optimization of the board furnish for a given end use.

437216 Greenhouse Gas Emissions of Ethanol Production from Lignocellulosic Biomass in Biorefinery System

Abstract:

Greenhouse gas (GHG) emissions reduction is one of the main drivers for bioenergy development to replace petroleum-based fuels and chemicals. Biorefinery concept has recently attracted the attention from governments and private sectors since this implies the effective and sustainable use of biomass resources for ensuring food and energy security, mitigating climate change, and meeting the demand for renewable based chemicals and materials. This objective of this work is to estimate the GHG emission of a biorefinery system to produce ethanol as a base case study. The different generations (first, second, third generation) and types (hardwood and corn stover) of feedstocks were compared. In addition, the amount of harvesting residue left on the field was examined considering fertilizer requirement. Two conversion processes such as biochemical and gasification process were simulated and compared with gasification option to produce electricity. The experimental lab results together with existing literature data were used for the GREET modeling work.

Briquetting characteristics of woody and herbaceous mixed feedstocks

Jaya Shankar Tumuluru, Ty Dansie and Craig C Conner
Biofuels and Renewable Energy Technologies Department, 750 University Blvd, Energy Systems Laboratory
Idaho National Laboratory, Idaho Falls, ID, 83415

www.inl.gov

Idaho National Laboratory

2015 AIChE Annual Meeting, Salt Lake City, November 8-13th, 2015

Idaho National Laboratory

Why biomass Briquetting

a) Higher mass per unit volume, b) less storage footprint, c) improved storage stability (reduces the dry matter losses), d) higher transportation efficiencies, e) makes biomass into a commodity like product

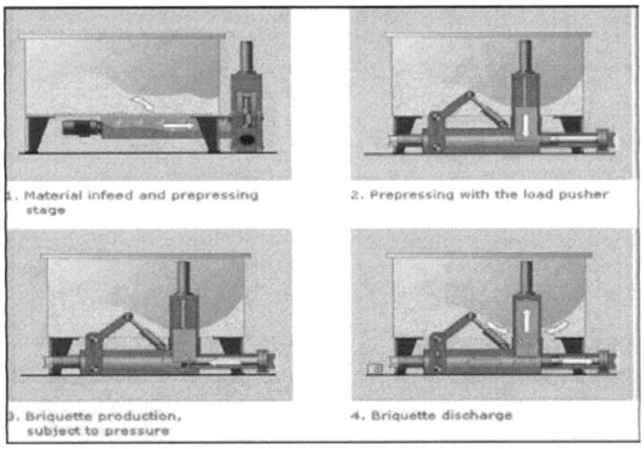

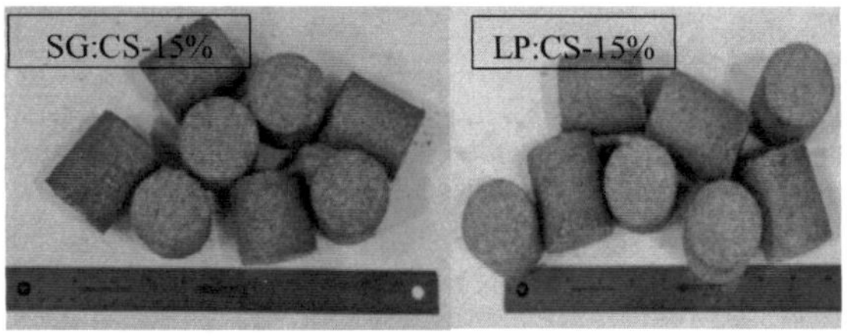

Ground biomass 60-80 kg/m^3: Briquettes 400-450 kg/m^3

Mixed feedstocks (50:50)		
Lodgepole pine/Switchgrass; Switchgrass/Corn stover; Corn stover/Lodgepole pine	12, 15, 18	4.8 & 12.7
Product properties	Briquette moisture content; Briquette unit, and bulk density; Durability rating and specific energy consumption	
Equipment	Hydraulic briquette press	

Unit and bulk density (kg/m3)

Idaho National Laboratory

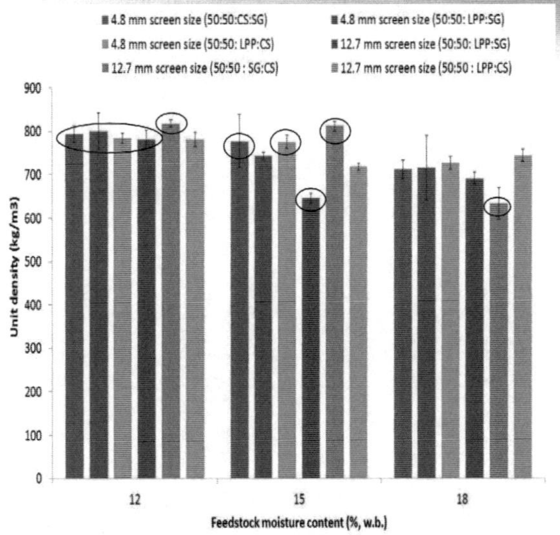

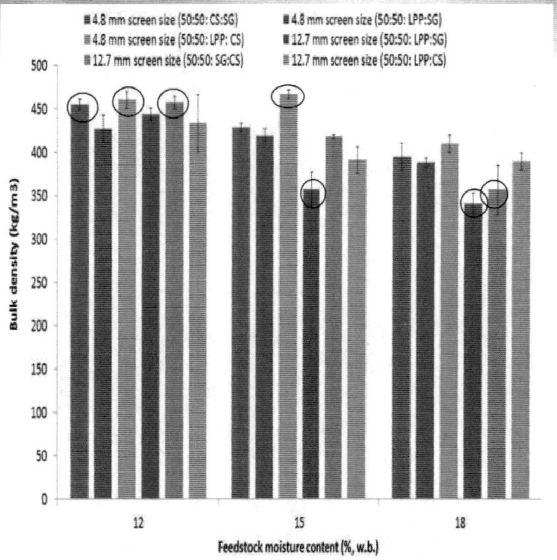

	Physical property		Process conditions	
Mixed Feedstock (50:50)	Unit density (kg/m3)	Bulk density (kg/m3)	Feedstock moisture content (%, w.b.)	Hammer mill screen size (mm)
SG:CS	777-825	400-450	12 & 15	4.8 & 12.7
CS:LP	775-783	433-466	12 & 15	4.8 & 12.7
LP:SG	782-801	419-444	12	4.8 & 12.7

Durability rating (%)

Idaho National Laboratory

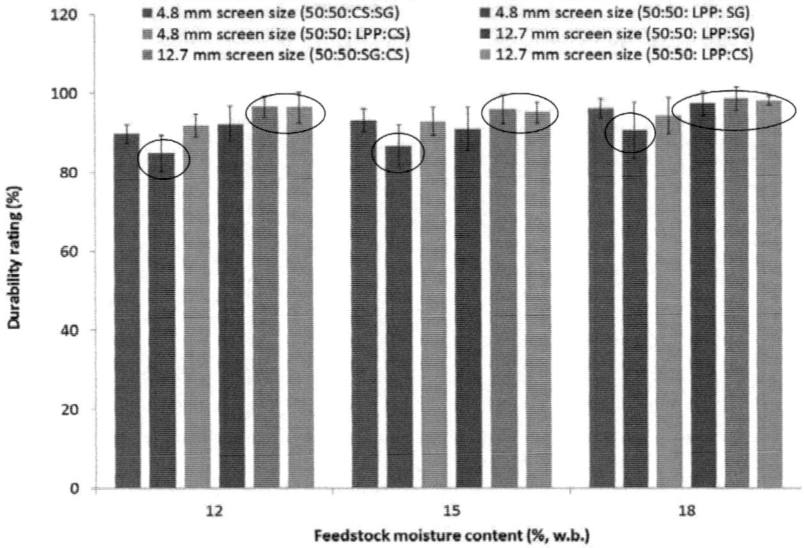

Feedstock	Physical property	Process conditions	
	Durability rating (%)	Feedstock moisture content (%, w.b.)	Hammer mill screen size (mm)
SG:CS	96-98	12, 15, 18	12.7
CS:LP	95-97	12, 15, 18	12.7
LP:SG	97	18	12.7

Idaho National Laboratory

Energy consumption (kWhr/ton)

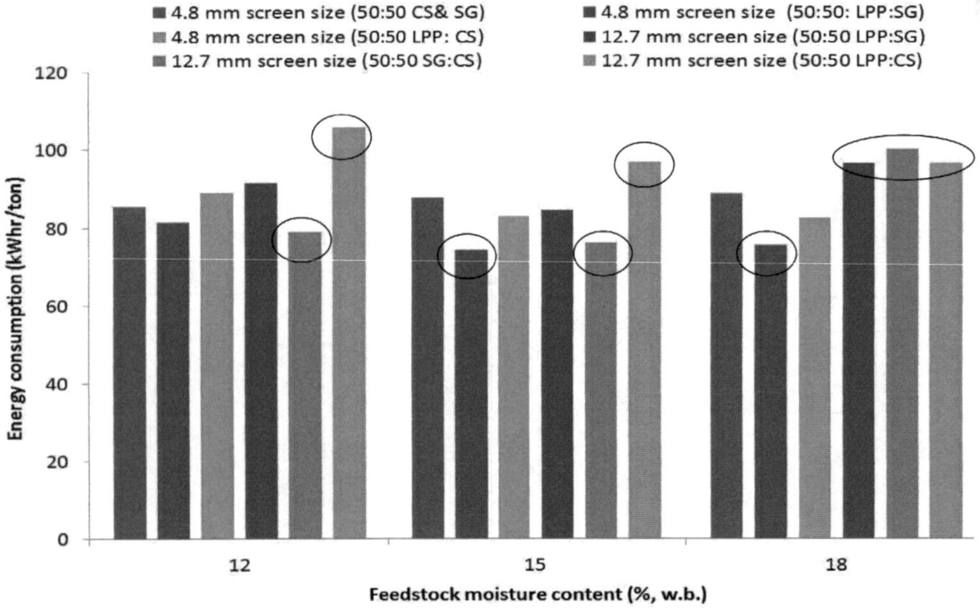

Feedstock		Process conditions	
	Energy consumption (kWhr/ton)	Feedstock moisture content (%, w.b.)	Screen size (mm)
SG:CS	76-78	12, 15	12.7
CS:LP	82-83	15, 18	4.8
LP:SG	74-75	15, 18	4.8

5

440271 Simulation and experiments of biomass pyrolysis for fuel gas with calcium oxide

Abstract:

In this study, the technical route of biomass pyrolysis in-situ CO_2 absorption with CaO has been proposed in order to get high-quality fuel gas. The optimum operation parameters were investigated with combination of experiments and simulating calculations, which will be the foundation of the large-scale application of this technology.

Through the simulations of thermodynamic equilibrium, the effects of temperature, pressure, mass ratio of CaO to sawdust (Ca/B) on the products distribution of the biomass pyrolysis in-situ CO_2 absorption were investigated. It indicated that atmospheric pressure, 700 and Ca/B of 0.65 would be the suitable operating conditions to produce the high quality fuel gas with good economic prospects. Based on the results of calculation, one system of a moving bed pyrolyzer coupled with a fluid bed combustor (MFS) has been used to produce high quality fuel gas. With the operating conditions of pyrolysis temperature of 700 , 0.65 of Ca/B and atmospheric pressure, the experiments were carried out in MFS system. The average contents of H_2 and CH_4 were about 40 vol.% and 26 vol.%, while those of CO and CO_2 were about 17 vol.% and 4 vol.%, respectively. LHV of the produced fuel gas was above 16 MJ/Nm^3. Cold gas efficiency of the system reached 51.2% through the analysis of system energy. Therefore, with high efficiency and simple operation, this technology may be a promising route to achieve high quality fuel gas for biomass utilization.

Effect of Fast Pyrolysis Conditions on Structural Transformation and Reactivity of Herbaceous Biomasses at High Temperatures

Anna Trubetskaya[1]
Markus Steibel[2]
Professor Hartmut Spliethoff[2]
Associate Professor Flemming Hofmann Larsen[3]
Senior Researcher Peter Arendt Jensen[1]
Professor Anker Degn Jensen[1]
Professor Peter Glarborg[1]

[1] DTU Chemical Engineering, Green research center, 2800 Lyngby, Denmark
[2] Department of Energy Systems, TU Munich, 85478, Garching, Germany
[3] Department of Food Science, University of Copenhagen

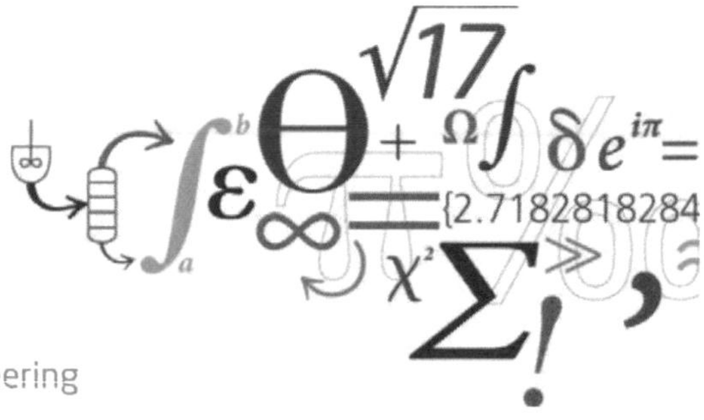

DTU Chemical Engineering
Department of Chemical and Biochemical Engineering

Objectives

- Herbaceous fuels (Danish wheat straw, rice husk)
- Pyrolysis of smaller particle size (0.2-0.4 mm)
- Experimental investigations at fast heating rates (10^4°C/s) and at high temperatures (up to 1500°C)
- Alkali and silicates effect on the char reactivity and morphology

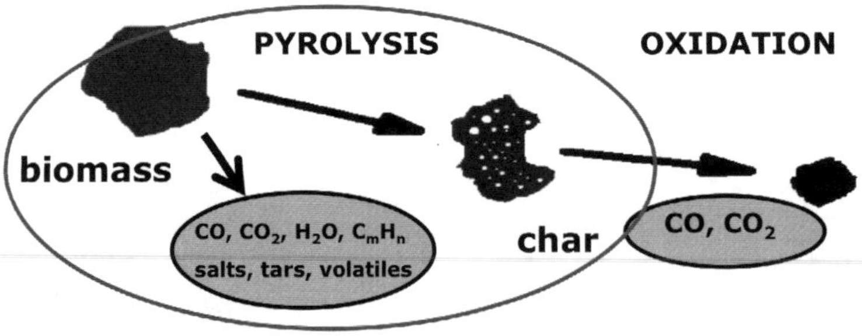

TU Munich Entrained Flow Reactor (BabiTER)

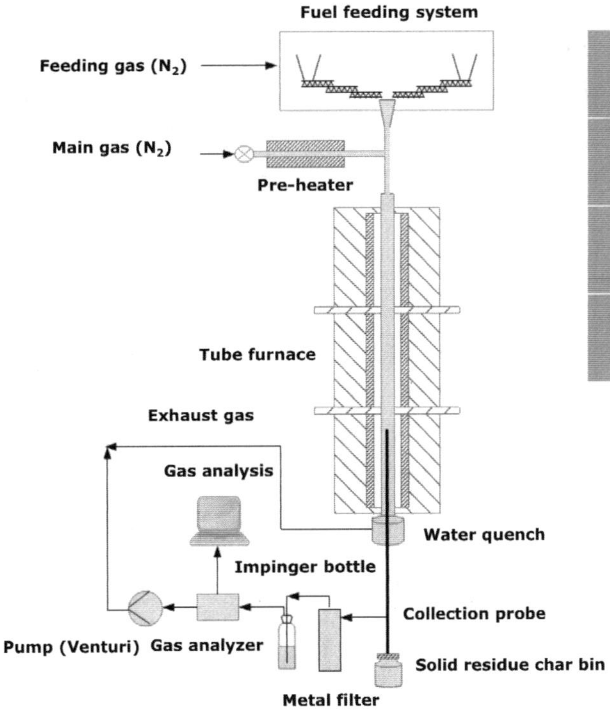

Atmosphere	N_2, O_2, H_2, CO_2,
$T_{maximal}$	1500°C
Heating rates	up to 10^4°C/s
Dimensions	L = 1.4 m D_{tube} =0.04m

Char morphology at 1000-1500°C

Rice husk

Wheat straw

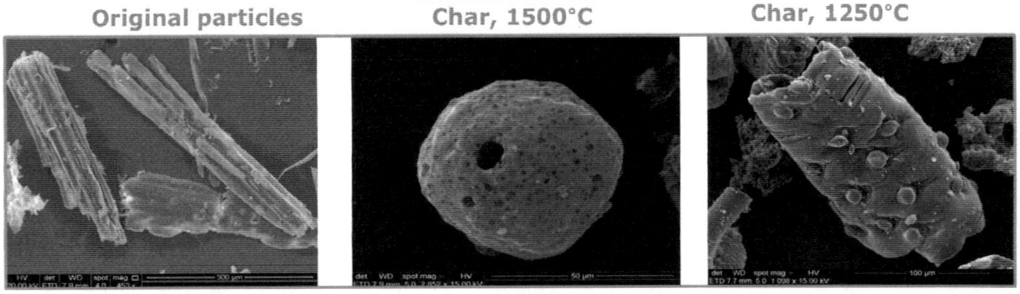

- The shape of the rice husk chars remained preserved
- Wheat straw chars obtained different shapes from near-spherical to cylindrical

Particle size analysis (CAMSIZER XT)

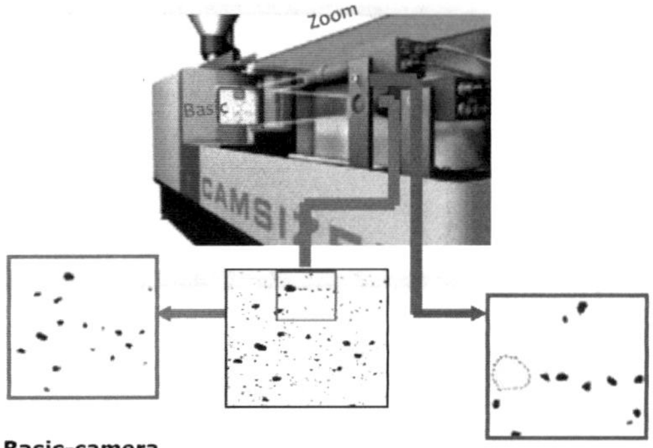

Basic-camera
(3 µm-30 mm)

Zoom-camera
(3 µm- 3mm)

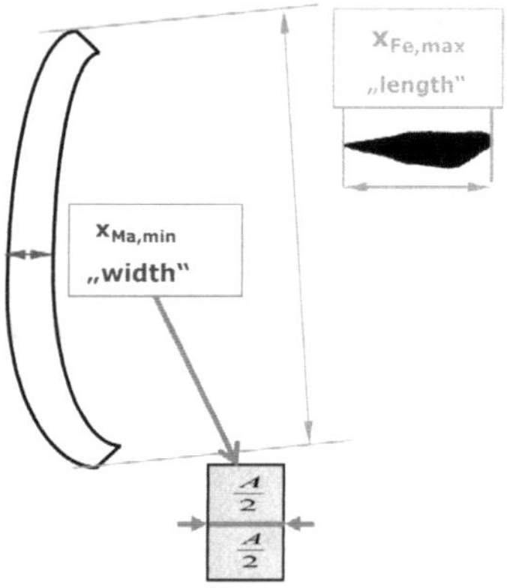

1. $\mathbf{x_{Ma,min}}$ - length of the line which bisects the particle ⟶ $q_3 = \frac{dQ_3(X_{Ma,min})}{x(x_{Ma,min})}$
2. $\mathbf{x_{Fe,max}}$ - distance between two maximal tangents directed parallel to each other

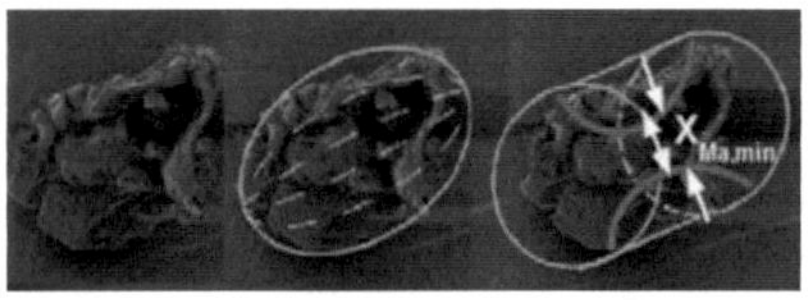

Particle shape analysis (CAMSIZER XT)

Biomass particle

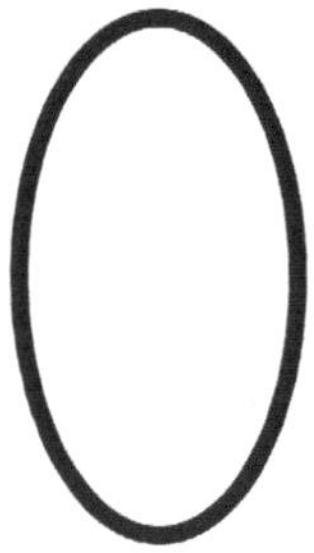

Ellipsoidic representation

1. Ratio width ($x_{Ma,min}$) to length ($x_{Fe,max}$) $\frac{b\ (width)}{l(length)}$

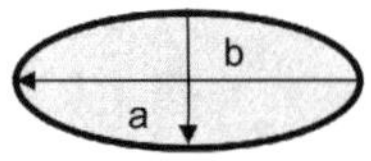

2. Sphericity (SPHT) $\frac{4\pi A}{P^2}$

Particle size analysis

Rice husk

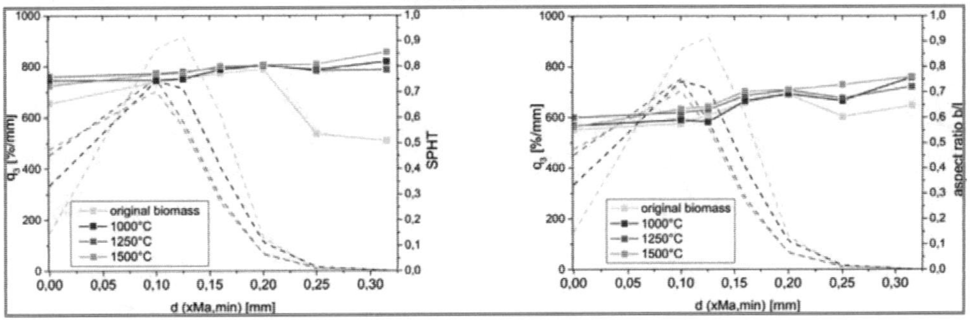

Wheat straw

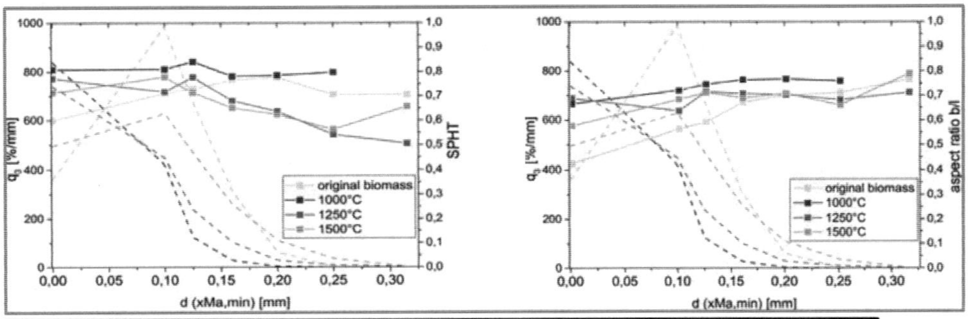

- Particle size of wheat straw chars was significantly reduced compared to the raw fuel
- Rice husk char kept the particle size of the original fuel

Ash compositional differences

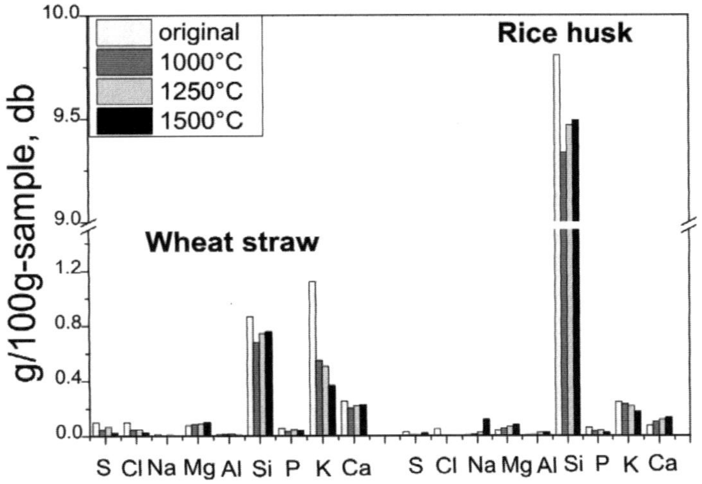

- The rice husk chars showed a high concentration of silicon oxides (> 90 wt.%) along with smaller amounts of potassium, aluminium, iron, sodium and magnesium
- The wheat straw chars contained potassium and calcium along with a high concentration of silicon

XRD analysis

DTU

Rice husk

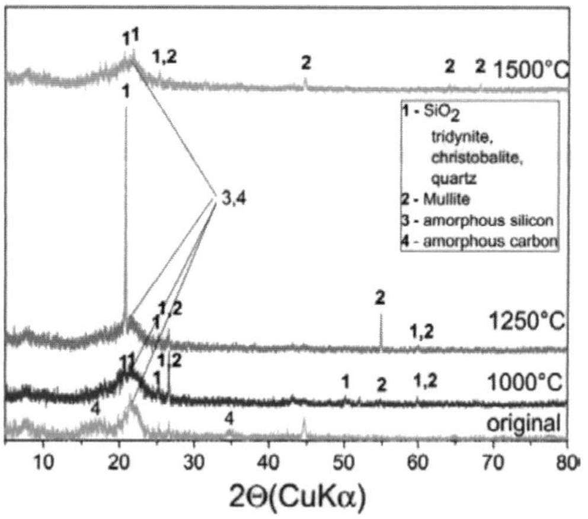

Wheat straw

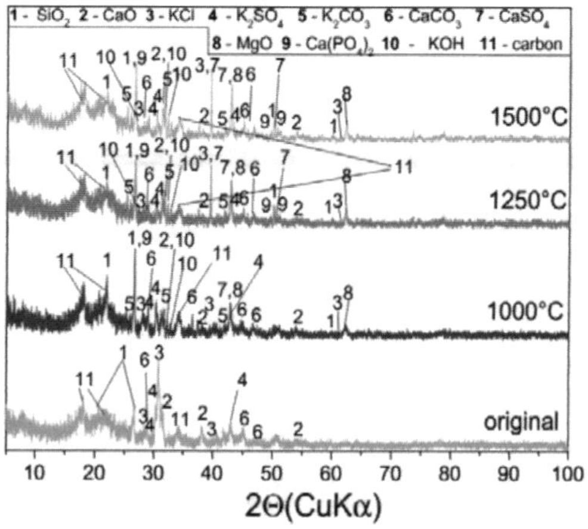

- The weak sharp and narrow reflections from crystalline silicon oxides as compared to the broad band of amorphous silicon oxides (2Θ= 21.8°) may indicate only partial crystallization under fast heating
- The XRD analysis of wheat straw char between 1000 and 1500°C showed a wide range of inorganic components, mostly present as oxides (SiO_2, CaO, MgO)

^{29}Si solid-state NMR analysis

DTU

^{29}Si CP/MAS and SP/MAS NMR spectra

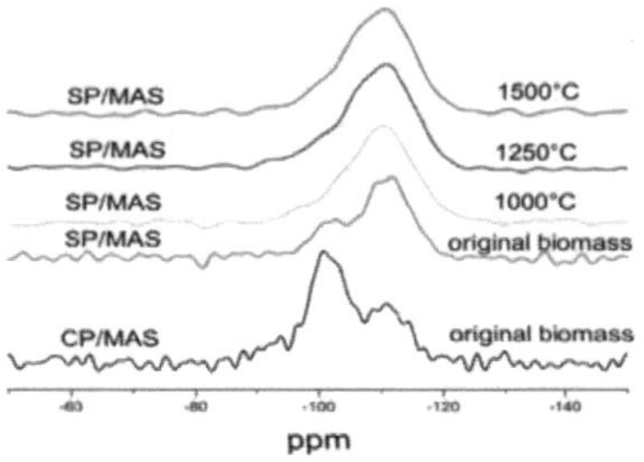

Original rice husk

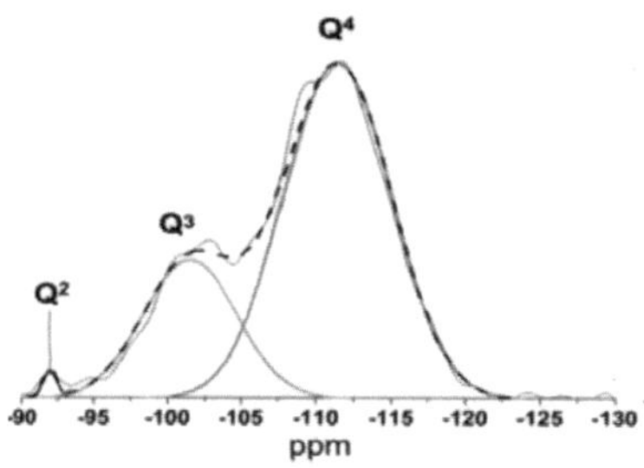

Char, 1500°C

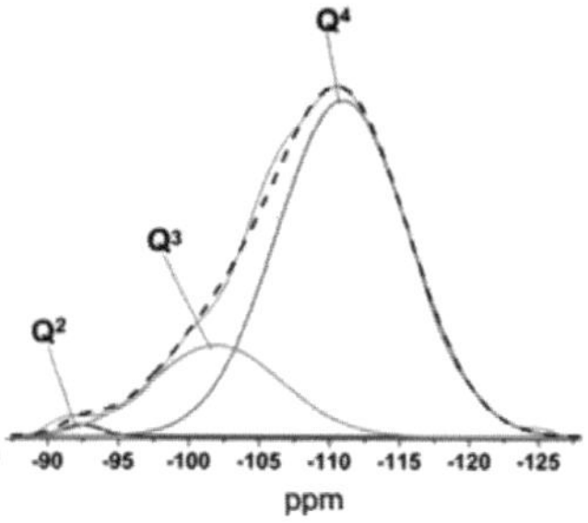

Q^{4*}, [Si(OSi)$_4$] : -111±3 Q^{3*}, [(OH)Si(OSi)$_3$]: -102±3 Q^{2*}, [(OH)$_2$Si(OSi)$_2$]: -92±5

* - index of Q^4, Q^3 and Q^2 indicates the number of siloxane bonds

- The silicon oxides were most likely present in the amorphous form due to the width of the resonances originating from the Q^4 sites
- No crystalline silicon compounds were observed
- No significant changes appeared in the ^{29}Si NMR spectra and thereby in the rice husk silicate structure with the increasing temperature
- The shape of the rice husk chars remained preserved even at a higher temperature, due to the low softening temperature of amorphous silicon oxides (≈730°C)*

Hedler et al. 2004

^{13}C solid-state NMR analysis

Rice husk and chars

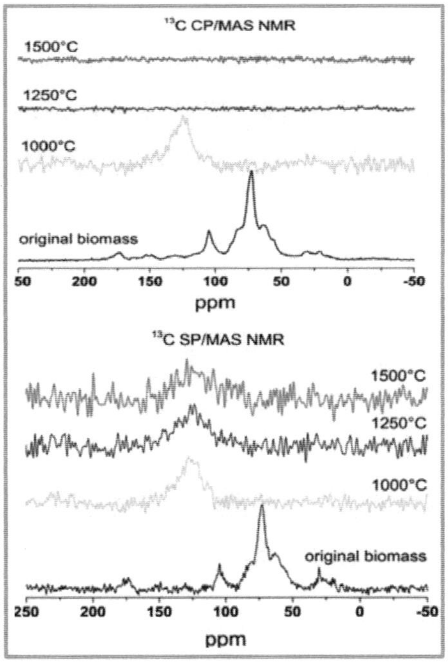

Original wheat straw and chars

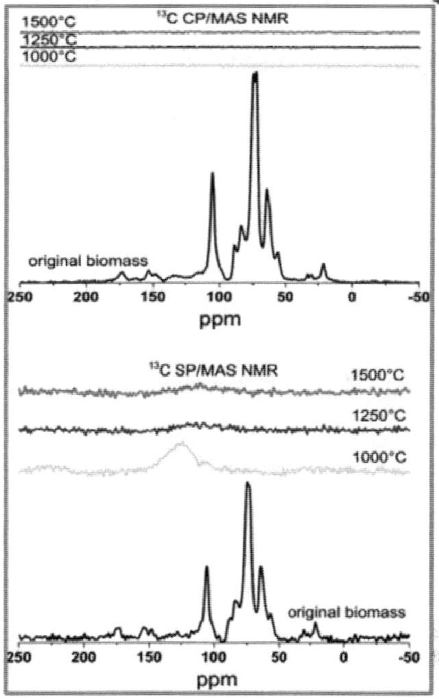

- No ^{29}Si CP/MAS NMR spectra of original wheat straw and its chars were observed compared to rice husk

Silicon oxides effect on char reactivity

TGA, 5 vol.% O_2 + 95 vol. % N_2

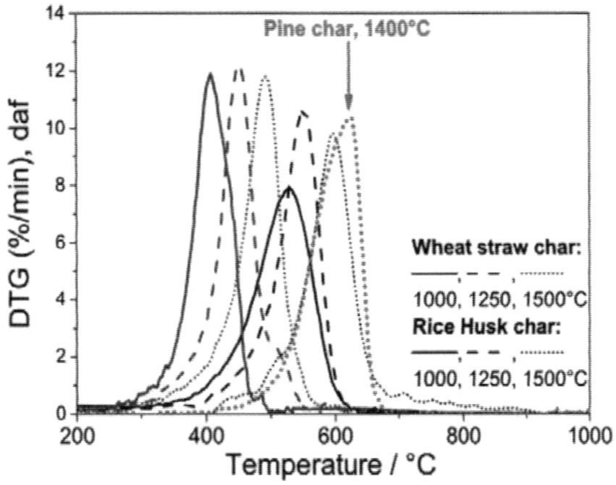

- Reactivities of pinewood and rice husk chars were similar in oxidation, indicating less influence of silica on the char reactivity
- The alkali rich wheat straw chars were 6, 18 and 50 times more reactive than wood and rice husk chars
- The reactivity of wheat straw and rice husk chars decreased at high heat treatment temperatures due to annealing of the carbon and loss of catalytically active potassium in the case of straw

Summary

- The results of the ^{29}Si solid-state NMR study indicated that the silicates in rice husk chars were mainly amorphous
- No significant changes appeared in the ^{29}Si solid-state NMR for Q^2, Q^3 and Q^4 branches of rice husk chars
- The presence of silicates in the rice husk affected the obtained char morphology by the formation of a glassy char shell due to the softening of the amorphous silicon oxides
- Silicates present in rice husk chars did not affect the reactivity in oxidation
- High alkali content in wheat straw increased the char reactivity

13 **DTU Chemical Engineering, Technical University of Denmark** November 11th, 2015

Acknowledgements

- Department of Energy Systems, Technical University of Munich
- Department of Food Sciences
- Dong Energy and Vattenfall
- Retsch Technology

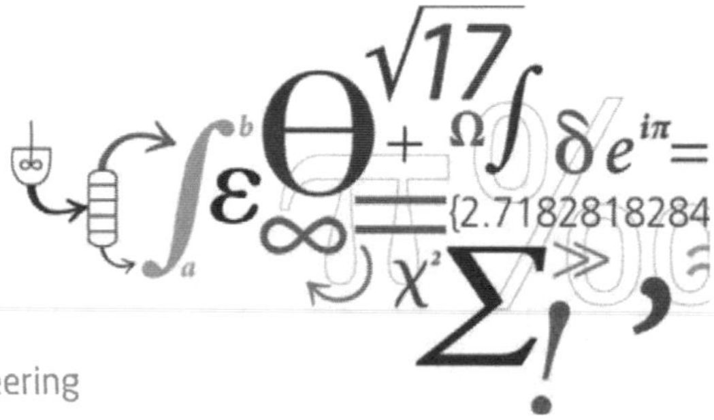

DTU Chemical Engineering
Department of Chemical and Biochemical Engineering

Thanks for attention!

Anna Trubetskaya
Technical University of Denmark
Chemical Engineering Department
Combustion and Harmful Emission Control Group
Søltofts Plads, Bygning 229
2800 Lyngby, Denmark
email: atru@kt.dtu.dk

DTU Chemical Engineering
Department of Chemical and Biochemical Engineering

433844 Structural Characterization of Biomass Derived Biochar Using Electron Energy Loss Spectroscopy

Abstract:

Biochar was previously considered as a low-valued solid fuel, which is the side product from biomass pyrolysis process. However, recent researches figured out the potential application of biochar as a renewable resource for selective adsorption medium, electrode material, and carbon fiber fillers. For the successful utilization of biochar, it is imperative to characterize the biochar structure. Many spectroscopic techniques, for example FT-IR, XPS, FT-Raman, NMR and XRD, partially revealed the structure. However, most spectroscopic techniques only provide qualitative information of specific functional groups in biochar. Without quantitative chemical bonding information, it is difficult to figure out the detailed structure of biochar. In this study, electron energy loss spectroscopy is used, which can quantitatively analyze chemical bonding states (sp^2 and sp^3 bonding) of biochar produced from loblolly pine under different temperature, atmosphere, and activation conditions. Two different gas atmospheres, nitrogen (100% nitrogen) and partial oxygen (7% oxygen + 93% nitrogen), were applied to different pyrolysis temperatures at 300°C, 350°C, 500°C, and 700°C. In addition, NaOH activation was applied to the selected biochar samples. Electron energy loss spectra analysis showed biochar carbon sp^2 bonding percent as 54.69% for O-300 (partial oxygen atmosphere and 300°C pyrolysis temperature) sample and 75.48% for O-700 (partial oxygen atmosphere and 700°C pyrolysis temperature) sample. The relationship between biochar structure and the production conditions will be presented in detail and its potential application will be discussed.

437229 Vibrational Characterization of Hydrothermal Chars: Simulation and Experiment

Abstract:

Carbonaceous materials derived from the thermal carbonization of carbohydrates and similar molecules in aqueous solutions under autothermal pressure have been shown to have potential applications as catalysts, energy storage materials, and water purification adsorbents. In the literature, char materials synthesized under liquid water conditions are termed "hydrothermal chars"; relative to pyrolysis-derived biochars, hydrothermal chars less highly aromatic and more highly oxygenated.

Despite on-going effort, improved characterization methods are required to develop rigorous structure-property relationships to guide design of hydrothermal chars for specific applications. Many techniques appear in the literature, ranging from elemental analysis (primarily combustion), various functional group identification (vibrational spectroscopy techniques and NMR), and extended structure characterization (primarily C-XANES). To date, infrared and Raman spectroscopy have been used strictly for functional group identification. However, the specific locations and intensities of vibrational absorption bands can provide a much richer understanding. Fully mining the vibrational data sets is limited by the complexity of the spectra, and this is the challenge that we are attempting to address.

In this work, we have used density functional theory, the available literature of model compounds and carbonaceous solids, and our own vibrational spectroscopy and microscopy for hydrothermal chars as complementary tools for understanding complex vibrational spectra. Specifically, we have modeled the Raman and infrared spectra of polyaromatic hydrocarbons and functionalized aromatic compounds. We can then view the individual vibrational modes of these model compounds to get a better understanding of how complex molecular structure is represented in Raman and IR spectra. Specific emphasis has been placed on understanding the molecular significance of the G and D bands, graphite and defect assignments based on spectroscopy of low-defect graphitic materials, for the poorly ordered and defect-rich hydrothermal char. In simple polyaromatics the wavenumber position of these bands was found to decrease with increase in the number of aromatic rings. The relative intensities of the characteristic peaks in the D and G band bands were found to fluctuate based on the number of rings, and the symmetry of the molecule. That is, whether the molecule consisted of rings arranged in the shape of simple polygons such as pyrene, or merely as a straight chain of rings as in tetracene. For example, molecules consisting of polygonal structures have decreasing wavenumber location of their D band, an A1G ring breathing mode, with increasing molecular size. Similar trends have been discovered for more complex aromatic compounds such as arylfurans and aromatic carboxylic acids. In addition, defects in the aromatic rings have been investigated including methyl and aldehyde side chains. The primary vibrational mode associated with the methyl group appears at 1465 +/- 10 cm^{-1}, and is largely independent of the number of rings in the aromatic structure. In terms of the influence of the methyl group on the characteristic aromatic vibrations, the effect is prominent for condensed aromatic structures with 1, 2, and 3 rings and then decreases for larger aromatic structures. The DFT simulation results were then used to examine experimental data for glucose and xylose hydrothermal chars.

437286 Biomass Pyrolysis for Closing Life Support Loops in Space

Abstract:

Manned spaceflight outside of low-Earth orbit will require significant advances in closing loops within life support systems, especially the recycling of solid and liquid wastes to produce oxygen, food, and fresh water. Within NASA's Technology Area 06 Human Health, Life Support & Habitation Systems Roadmap, waste management, water recovery and management, and habitation are three of the major Environmental Control & Life Support Systems and Habitation Systems functions. Long-term (2011-2029) technology development goals for these functions include recovering more water from more waste streams, transitioning from waste stabilization and reduction to material recovery, and enabling food production. Pyrolysis, or heating in a limited oxygen environment, is one waste conversion method that has received considerable attention due to its potential to enable improvements in all three functions. Here, we explore the development of a moderate-temperature slow pyrolysis reaction system that can transform solid waste and brine from the water treatment system into a nutrient-rich crop growth medium, while recovering water and carbon dioxide.

437308 Evaluation of Sugars and Bio-Oil Production Using Switchgrass Feedstock Grown in Heavy Metal Contaminated Soil

Abstract:

A combination of biochemical and thermochemical process to produce sugars and bio-oil by using switchgrass which was grown in lead contaminated soil was studied. Four different process routes which involve fast pyrolysis, enzymatic hydrolysis, acid hydrolysis, and their combinations were investigated. The first route only involved fast pyrolysis process only. The second route involved an acid hydrolysis pretreatment followed by the fast pyrolysis. In the third route biomass was hydrolyzed enzymatically to produce sugars. The solid remaining then was pyrolyzed to produce bio-oil. In the fourth route, biomass was pretreated with acid hydrolysis prior to the enzymatic hydrolysis to produce sugar followed by pyrolysis of the remaining solid to produce bio-oil. The impact of lead content in the switchgrass feedstock used for the production of sugars and bio-oil along with the recovery of lead through the above mentioned routes were evaluated by using switchgrass grown in regular soil for comparison. The yields of sugars from acid hydrolysis and enzymatic hydrolysis did not seem to be affected by the lead contained in the switchgrass. The chemical product distribution in bio-oil produced from pyrolysis of lead-containing switchgrass and regular switchgrass showed slight differences. As expected, acid hydrolysis and enzymatic hydrolysis prior to fast pyrolysis affected the product distribution by increasing its selectivity towards aromatics and ketones. Aromatic production was increased by 10% and ketones by 5%, while reducing phenols by 10%. The highest overall yields of the liquid products were obtained from the fourth route which utilized all the three process steps, resulting to the lowest bio-char byproduct. On lead recovery, it was found that most of the lead was extracted from biomass into the liquids in the hydrolysis steps, which can later be recovered by precipitation to for salts. Pyrolysis of the solid remaining from the combined both hydrolysis steps resulted in the bio-char which had the lowest content lead (2ppm) which is not believed to be dangerous by EPA standards

433764 13C NMR Spectroscopic Investigation of the Effects of Functional Groups on Catalytic Pyrolysis Oil Properties

Abstract:

13C NMR spectroscopic investigation of the effects of functional groups on catalytic pyrolysis oil properties.

Ofei D. Mante[a] and F. A. Agblevor [b],

[a]Department of Sustainable Energy Technologies, Brookhaven National Laboratory, Upton, NY

[b]USTAR Bioenergy Center, Department of Biological Engineering, Utah State University, Logan, UT

Abstract

The production of infrastructure-ready biocrude oil from catalytic pyrolysis offers an opportunity to co-process biomass derived intermediates alongside petroleum feedstocks such as atmospheric gas oil and vacuum gas oil in standard refinery units. Recent studies suggest that, up to 20wt% of upgraded bio-oils can be directly cracked in fluid catalytic cracking (FCC) process with these petroleum feedstocks without affecting the yields. Nevertheless, one of the critical success factors for commercial scale bio-oil co-processing is the physicochemical properties of the upgraded bio-oil. Since the upgrade bio-oil produced in catalytic pyrolysis is not completely free of oxygenated species, there is the need to fundamentally understand the impact of its chemical composition on the physicochemical properties. In this work, several upgraded bio-oils with varying fuel qualities produced from catalytic pyrolysis of different biomass feedstocks with HZSM-5 were characterized. Using ^{13}C NMR spectroscopy, we were able to gain insights on the effect of the various functional groups on the physicochemical fuel properties: long-term storage stability, acidity, viscosity, density, solid residue, and elemental composition. The evidence from this investigation suggests that residual oxygenated species such as phenolics, anhydrosugars, and carbonyls continue to negatively impact the fuel properties of even upgraded bio-oils with low oxygen content levels (e.g.,12 wt.%.). In this talk, we will also discuss the opportunities and challenges of utilizing biocrude oil as a feed for standard refinery units.

427509 Controlling Phase Stability of Biomass Fast Pyrolysis Bio-Oils

Abstract:

AIChE Annual Meeting 2015, T4A08 Thermochemical Conversion of Biomass I

Controlling Phase Stability of Biomass Fast Pyrolysis Bio-oils

Anja Oasmaa, Tom Sundqvist, Eeva Kuoppala, Manuel Garcia-Perez, Yrjö Solantausta, Christian Lindfors

Fast pyrolysis bio-oil is entering into the heating oil market to replace fossil fuel oils. First commercial size pyrolysis plant is under commissioning in Joensuu, Finland, by Fortum and the next one is under commissioning by Empyro in the Netherlands. Large-scale combustion of bio-oil has been demonstrated and standards and norms are being prepared. Information on critical fuel oil properties related to bio-oil combustion is available. However, there is insufficient data on the phase stability of bio-oils and especially how to predict and prevent phase-separation. It is important to control this property because phase-separated bio-oil will cause e.g. pumping problems, blockages, and irregular combustion.

Fast pyrolysis bio-oil is a chemically diverse mixture consisting of thermal degradation products of lignocellulose. The complexity of chemical compounds in bio-oil varies from simple oxygenated C1 compounds to high molecular weight oligomers. Bio-oil contains oxygenated functionalities, such as hydroxyl, carbonyl and carboxyl in both aliphatic and aromatic structures. The homogeneity of bio-oil depends on the ability of the dissimilar molecules to dissolve each other. The water-soluble and water-insoluble compounds in bio-oil can either be in one homogeneous phase or form two individual phases, to which we refer to as phase separation. Phase separation can occur immediately after condensation of the pyrolysis vapours to bio-oil due to certain pyrolysis conditions or type of raw material, or after years of ageing because of changes in composition caused by re-polymerization reactions.

In this study we present how the phase separation of fast pyrolysis bio-oils is related to the chemical composition, and show that the probability of phase separation can be predicted with a numerical stability index based on the chemical composition. The chemical composition of the bio-oils studied was characterized using a solvent extraction scheme that describes the composition of bio-oil as a blend of three macro fractions: C1-C6 oxygenated molecules (named co-solvents), water-insoluble molecules and water-soluble polar molecules (including water but excluding the co-solvents), e.g. anhydrosugars. The results show that the required amount of the co-solvent to dissolve both fractions and keep the oil homogeneous varies depending on the chemical composition. The minimum amount of co-solvent for homogeneous bio-oils was observed to be from 15 to 30 wt% depending on the composition of bio-oil. The co-solvent content in fresh bio-oils was typically from 20 to 30 wt% and decreased to as low as 5 wt% after prolonged ageing. Correlation of chemical composition and homogeneity of fresh and aged bio-oils are shown in ternary phase diagrams. Addition experiments were made with model compounds to cover a larger part of the phase diagram. The model compounds were water, a model solvent mixture and an isolated water-insoluble fraction. The numerical stability index was generated based on the ternary phase diagram.

419098 Liquid Hot Water Pretreatment of Lignocellulosic Biomass to Enhance Sugar Recovery and Enzymatic Digestibility of Cellulose

Abstract:

Several processes are required in cellulosic ethanol production from lignocellulosic feedstocks such as pretreatment, enzymatic hydrolysis, fermentation, and product separation/ distillation. The native lignocellulosic biomass has limited its accessibility to enzymes and microorganisms due to its complex cell wall structure of cellulose-hemicellulose-lignin. Therefore, pretreatment is an essential prerequisite to overcome recalcitrance of biomass and enhance the bio-chemical conversion ratio of polysaccharides.An effective pretreatment can reduce the downstream pressure by making cellulose more accessible to the enzymes and minimize the formation of degradation products that inhibit the growth of fermentative microorganisms. For the advantages of without additional chemicals, less inhibitory products and higher pentosan recovery, liquid hot water pretreatment become attractive.

The process of step-change flow rate (SCF) liquid hot water pretreatment and step-change temperature (SCT) liquid hot water pretreatment were developed in GIEC, respectively, for the sweet sorghum bagasse (SSB) and eucalyptus wood chips (EWC) to enhance sugar recovery and enzymatic digestibility of cellulose. Furthermore, the decomposition mechanism of hemicellulose and lignin in the liquid hot water was investigated.

Moreover, different structural changes at the plant tissue, cellular, and cell wall levels were investigated to understand the decomposition mechanism of sugarcane bagasse cell wall in the liquid hot water pretreatment. The transmission electron microscopy (TEM) images showed that sugarcane bagasse cell walls were composed of the middle lamella (ML) layers, the primary wall (P) layers, and the secondary wall layers. While after the pretreatment, the boundaries among the S1, S2 and S3 layers in the secondary wall of treated samples could not be distinguished exactly. The data from the scanning electron microscopy and energy dispersive X-ray analysis(SEM-EDXA)shows that the migration of lignin happens among different cell wall layers. Moreover, pseudo-lignin, the degradation products of lignin and xylan, appeared on the surface of pretreated sugarcane bagasse. Furthermore, Raman spectra of treated sugarcane bagasse indicated that the distribution of cellulose in the cell wall was homogenized, and the difference in the chemical composition was reduced. All of these changes on the ultrastructural level of the cell wall were responsible for the improvement of the enzymatic digestibility of treated samples.

417409 An Aggregated Understanding of Alkaline Hydrogen Peroxide (AHP) Conversion of Biomass Hemicelluloses to Organic Acids

Abstract:

Hemicellulose represents up to 30% of biomass dry weight been the second most abundant non-food carbohydrate only next to cellulose. However, due to its lower degree of polymerization and crystallinity compared to cellulose, hemicelluloses are much more susceptible to degradation during chemical and thermal treatment. The ensuing degradation products have little value and can even be detrimental for subsequent fermentation and downstream processing. Designing a pretreatment chemistry to guide hemicellulose degradation toward the formation of useful products is of significant interest. We recently demonstrated that hemicellulose can be converted to carboxylic acids by AHP treatment, especially in the presence of a metal catalyst. This provides a potential and appealing pathway for the conversion of hemicellulose to high value compounds. However, a detailed understanding of the mechanisms involved in metal catalyzed AHP reactions with hemicellulose and the effects of metal catalysts on reaction kinetic and yields have not been studied before.

This study investigated the interaction of peroxide reactive species with hemicellulose, its role on the conversion to carboxylic acids, and describes the role of metal catalysts on the generation of beneficial reactive species for hemicellulose conversion. Due to the complexity of the process, the first approach was to identify key peroxide derived reactive species responsible for conversion using model hemicellulose compounds (i.e. glucomannan and xylan). The metal catalyst interaction with reactive species was investigated to improve the efficiency of hemicellulose conversion using both metal ions and organometallic complexes. Finally, we develop a deeper understanding of the reaction mechanism and reaction kinetics using a simplistic model that describes three main reaction pathways taking place during conversion. From these studies, we have been able to describe the effect of metal ions, organometallic complexes and single reagents on AHP. We believe that the incorporation of hemicellulose-to-carboxylic acids as an additional product without the need of additional steps during biomass pretreatment could be determinant into inclining the balance toward a successful biorefinery industry.

419041 Effect of Lignin Content on Hemicellulose Hydrolysis during Liquid Hot Water Pretreatment

Abstract:

Recalcitrance to sugar release is a major limitation for cost-effective industrial conversion of lignocellulosic biomass to biofules. Lignin layers surrounding hydrophilic hemicelluloses and cellulose fibrils in the plant cell walls protect them from the chemical and enzymatic attack. Many work had been reported about the contribution of lignin removal to cellulose accessibility to cellulose, but less to the hemicellulose hydrolysis in the pretreatment process. In this paper, the effect of lignin content on hemicellulose saccharification during liquid hot water pretreatment was investigated. Compared with pure birch xylan, the samples with 50% added kraft lignin showed a yield increase of xylose, xylobiose, xylotriose and xylotetraose with 6,4,2.2 and 1.5 times, respectively, after the liquid hot water pretreatment of 180 °C for 30 min. Similarly, the yield of total xylose decreased with the lignin removal from sugarcane bagasse with peracetic acid. Furthermore, there hybrid energy sorghums with same xylan content but different lignin content were selected to evaluate the effects of lignin content on hemicellulose hydrolysis. The data showed that there was lower initial rate of hemicellulose hydrolysis for the sorghum with high lignin composition. At the same time, the rate of xylose degradation was reduced due to the protection of lignin. All of these results indicated that lignin content was an important factor for the optimization of liquid hot water process and studies on plant cell wall decomposition mechanism.

437030 Hydrolysis Kinetics of Woody Biomass: Surface Renewal

Abstract:

Woody biomass is known to be composed of three major components: cellulose, lignin and hemicellulose. Cellulose is the most stable natural polymer in woody biomass, a linear dehydration polymer of b-D-glucose. Glucose is the common sugar that has been the substrate in ethanol production and other fine chemicals since the early humanity. This is naturally the sugar of importance in a biorefinery. To get to the glucose or cellulose, one must deal with the other components of the woody biomass. Lignin is a macromolecular substance of phenopropane units. It has the highest energy density among all the components. Hemicellulose is heteropolymer of five and six carbon sugars with other residues such as acetyl. Hemicellulose is the easiest to degrade among the major woody biomass components. Hydrolysis of woody biomass produces Xylose, Galactose, Glucose, Mannose, Rhamnose and Arabinose, along with their oligomers. In this presentation, we show that the mechanism of woody biomass hydrolysis occurs from the solid surface, and the kinetics follow closely to the surface renewal model. The apparent order of reaction is zero. The kinetic model agrees quite well with hot-water extraction of woody biomass.

437225 Gasoline Production By One-Pot Catalytic Conversion of Lignocellulosic Biomass Derived Sugar/Polyol

Abstract:

Gasoline production by one-pot catalytic conversion of biomass derived sugar/polyol was investigated over bifunctional heterogeneous catalyst. Ni@HZSM-5/SG catalyst showed excellent performance of converting sugar/polyol to gasoline. The gasoline yield of 46.94% was obtained. Its composition and properties are similar to petroleum based gasoline. It provides a novel approach for the production of gasoline by one-pot aqueous phase catalytic conversion of sugar/polyol derived from biomass.

437228 Micro-Aerobic Fermentative Hydrogen Production from Corn Stalk

Abstract:

To verify the feasibility of micro-aerobic fermentative hydrogen production from straw, micro-aerobic and anaerobic fermentative hydrogen production was carried out in a bottle test (150 mL) and scale-up experiment (10 L) using untreated and acid- and alkaline-pretreated corn stalk (CS) as substrates. A maximum increase of 68.71% of hydrogen production was obtained in the bottle test with addition of 280 mL/(kg total solids [TS]·day) oxygen from acid-pretreated CS. Micro-aerobic fermentation increased the hydrogen yield and production rate by 43.3% and 42.3%, respectively, in scale-up experiments compared to anaerobic fermentation. Micro-aerobic fermentative hydrogen production was conducted at an oxidation reduction potential of -300 to -150 mV. Both micro-aerobic and anaerobic hydrogen production were butyric acid-type fermentation.

437322 Integrated Technology for Liquid Biofuels and Bioproducts from Lignocellulose

Abstract:

With lignocellulose as a raw material, based on separation of components and total utilization of lignocellulose biomass, the pretreatment of lignocellulose was studied. After pretreatment, cellullase production by a novel solid state fermentation process was investigated. Then, cellulose was hydrolyzed by cellulase to get hydrolysate which was used to produce bioethanol by simultaneous saccharification and fermentation. In order to reduce the inhibition of bioethanol on fermentation, CO_2 gas stripping method was adopted and discussed. In addition, hemicellulose was used to produce bioethanol, 2,3-butanediol and xylitol, respectively. As to lignin, it was used to produce lignin derived products. Furthermore, on the one hand, bioethanol can be used as biofuel; on the other hand, bioethanol can be used to produce bioethylene by catalytic dehydration which was also studied in our research work. As we know, ethylene is a basic raw material to be used to produce more than 70% chemicals in chemical industry. In addition, biolipid production by fermentation to be used as raw material to produce biodiesel was also investigated, and biodiesel production by chemical transesterification with solid catalyst and supercritical fluid technology respectively were discussed too. Moreover, biobutanol production by fermentation was studied in our research. Therefore, an integrated technology route of bioenergy and biorefinery from lignocellulose were presented by our studies.

430875 Production of High Quality Bio-Oil Via Fast Co-Pyrolysis (FCP) of Cellulose and Polypropylene

Abstract:

Production of High Quality Bio-oil via Fast Co-Pyrolysis (FCP) of Cellulose and Polypropylene

Deepak Kumar Ojha and R. Vinu

Department of Chemical Engineering and National Center for Combustion Research and Development, Indian Institute of Technology Madras, Chennai- 600036, India

E-mail: deepakojha86@gmail.com, vinu@iitm.ac.in

Abstract

Fast pyrolysis has received a great attention for direct conversion of biomass into liquid fuels and valuable intermediates. The bio-oil produced from the process has high oxygen content and needs further upgradation to make it usable as a transportation fuel. Currently this is achieved via catalytic hydrodeoxygenation which is highly energy and hydrogen intensive. Co-processing of biomass with polymers is a novel strategy and this is expected to increase the product quality in terms of H/C ratio and heating value, thus reducing the upgradation cost. Detailed understanding of these reactions is prerequisite for the technological development and commercial implementation of the process. Fast co-pyrolysis (FCP) is thought to proceed via complex network of competing free radical and concerted reactions such as aldol condensation, retro-Diels-Alder reaction, β-scission and hydrogen abstraction between biomass and polymer free radicals.

In this study, fast co-pyrolysis of cellulose and polypropylene (PP) is investigated. The objectives of this work are two fold: firstly, the effects of cellulose: PP ratio on the yield of individual products and heating values were evaluated at different temperatures from 400 to 700 °C. Secondly, the evolution of major functional groups in a short time period of a few seconds is studied. Fast co-pyrolysis experiments were conducted in a Pyroprobe® 5200 pyrolyzer (CDS Analytical Inc.) which was interfaced with multidimensional gas chromatograph/mass spectrometer (2D-GC/MS, Agilent Technologies). For the real time function group analysis, the Pyroprobe was interfaced with Cary 660 Fourier transform infrared spectrometer (FT-IR, Agilent Technologies). ***Table 1*** depicts the yields of major products formed during fast co-pyrolysis of cellulose and PP of different compositions. Significant difference between calculated and experimental product yields shows that interactions between cellulose and PP are manifested in the form of alcohols. Importantly, the alcohols belonged to C10-C15 carbon number range. Specifically, long chain alcohols like (2,4,6-trimethylcyclohexyl) methanol, 2-isopropyl-5-methyl-1-heptanol and 11-methyldodecanol were obtained in significant amounts when PP content was high (50-75 wt.%) in the mixture. The heating values of the products were calculated using the empirical relation of Davenport, and interestingly, the heating values correlated well with the sum of yields of alcohols, hydrocarbons and monoaromatics. Overall, two fold improvement in heating value of the condensable products was observed. Higher pyrolysis temperature led to the formation of mono and poly aromatic hydrocarbons. The real time functional group analysis indicated a reduction in reaction time with an increase in PP content in the feed. The time taken for the maximum yield of various products corresponded to 10 seconds, while pyrolysis was complete in less than a minute.

Table 1. Comparison between experimental and calculated composition of the major products formed in co-pyrolysis of cellulose and PP of different compositions at 500 °C.

Feed Composition	**C:PP 100:0**	**C:PP 75:25**		**C:PP 50:50**		**C:PP 25:75**		**C:PP 0:100**
Products	*Expt.*	*Expt.*	*Calc.*	*Expt.*	*Calc.*	*Expt.*	*Calc.*	*Expt.*
Alcohols	6.95	22.73	5.21	27.81	3.48	30.87	1.74	0
Hydrocarbons	0.67	28.23	23.56	40.21	46.44	43.65	69.33	92.21
Anhydrosugars	17.36	7.11	13.02	2.92	8.68	1.70	4.34	0
Aldehydes/Ketones	23.20	10.63	17.40	5.80	11.60	1.83	5.80	0
Furans	27.89	9.92	20.92	6.20	13.95	1.76	6.97	0
Char	5.67	11.29	6.20	14.32	6.73	13.28	7.27	7.8

428152 Advanced Fuels and Coproducts from Guayule and Algal Feedstocks Via Tail-Gas Reactive Pyrolysis (TGRP) and Subsequent Upgrading

Abstract:

Thermochemical conversion technologies (e.g. pyrolysis, gasification) are the only technologies capable of converting an entire plant into fuels compatible with refinery processes. Such technologies are also necessary to make use of excess refuse from plants, including sugar cane bagasse, oil palm residues, and sorghum straw to name a few. While pyrolysis is considered as a "feedstock-neutral" technology for producing liquid fuels, particular feedstocks are more problematic than others. Due to higher concentrations of non-carbohydrate components like proteins, fats, lignin, or even inorganic components, some feedstocks have vastly different behavior when exposed to pyrolysis processes, which in turn can sometimes restrict their implementation.

The tail-gas reactive pyrolysis (TGRP) process converts feedstocks which were otherwise impossible with traditional pyrolysis. Our talk will focus on two case studies. First, guayule bagasse normally cannot be pyrolyzed into usable oil due to the very high viscosity resins present – only a viscous solid is produced. With the TGRP process, low viscosity bio-oils are produced with as low as 8 wt% oxygen on a dry basis. Second, algae-based bio-oils are very heavily nitrogenated (>10 wt%) in addition to the usual oxygen content; also, heavy salts and high lipids content make pyrolysis processing difficult. Distillates of bio-oils from both feedstocks produced high yields of low molecular weight organics. For all cases, TGRP distillates enhanced the efficiency of heteroatom removal for production of fuels and commodity chemicals, compared with traditional bio-oils. Use of high space velocities (LHSV > 0.6) is possible when TGRP oil distillates undergo continuous hydrotreatment in a packed bed reactor. Due to the absence of coke precursors, HDO of distillates produces as high as 70 – 80 wt% yields of hydrocarbons. For algae-based bio-oils, a one-step simultaneous HDO and HDN was achieved, producing finished hydrocarbon fuel compounds.

430160 Hydrothermal Liquefaction: Effect of Various Waste Streams As Reaction Medium

Abstract:

Abstract

Hydrothermal liquefaction of biomass is the thermochemical conversion of biomass into liquid fuels by processing in a hot, pressurized water environment for sufficient time to break down the solid biopolymeric structure to mainly liquid components. Typical hydrothermal processing conditions are 250–374 °C of temperature and operating pressures from 40 to 220 bar. The process is meant to provide a sustainable treatment of wet materials without the need for drying and to exploit ionic reaction conditions by maintaining a liquid water processing medium. The temperature range is sufficient to initiate pyrolytic mechanisms in biopolymers while the pressure is sufficient to maintain a liquid water processing phase.

The current hurdles for HTL commercialization are the cost associated with the very high operating pressure as well as low and often unpredictable conversion efficiencies. A significant amount of research and development on catalytic HTL has been undertaken in order to provide higher yields. The most commonly considered "catalyst" has been different alkalis, which modify the ionic medium to favor base-catalyzed condensation reactions and lead to formation of aromatic oils. Dairy manure, sewage sludge from water treatment plant, and digestate from anaerobic digester are slightly basic mainly because of the presence of urine, alkaline solution, and ammonium hydroxide, respectively. Moreover, these wastes contain significant amount of inorganics, ranging between 10% and 40% based on their dry organic matter. While these wastes maybe considered as liability in the dairy and waste treatment industries, an alternative, value-added approach for processing the waste is presented in this work.

In this study, feasibility of waste streams e.g. dairy manure, sewage sludge etc. as reaction medium for HTL of pine wood (*pinus taeda*) was evaluated. HTL conditions for this study were limited to 250-300 °C and 5-30 min. The quality of the biocrude was determined by means of heating value, elemental analysis, viscosity, density, and GC-MS analysis. In the absence of catalysts, the biocrude yield was shown to increase with HTL time and temperature. The results also revealed an increase in the yield by addition of sewage sludge and dairy manure to the lignocellulosic biomass feedstock. GC-MS spectra pointed out that despite the similar energy value of the biocrude; the chemical constituents depend upon the reaction media. The presence of different inorganics in the reaction medium most probably had a catalytic effect, in turn altering the biocrude composition. These studies suggest that waste streams applied as reaction medium for HTL of woody biomass improve yield and reduce the reaction temperature (and pressure). The presentation will give a snapshot of the status of the ongoing experimental program, and the direction for future research.

417899 Chemical Modification of Hydrothermal Chars Using Mechanical Energy

Abstract:

Hydrothermal carbonization has potential for synthesis of materials with catalysis, water purification, and energy storage applications. Functionalization of the carbon surface will be required to increase effectiveness for a specific application. Mechanical methods have potential for chemical modification of hydrothermal chars with reduced environmental burdens and costs compared to traditional chemical methods. In this work, we produced hydrothermal char by heating a glucose solution to 180°C at autothermal pressure. The synthesized hydrothermal chars were then subjected to ball milling at times ranging from 30 min to 5 hours. Structural changes were evaluated using a suite of methods. Combustion analysis methods were used to confirm that the hydrothermal char underwent some combination of dehydration, decarbonylation, and decarboxylation reactions during milling with the net result being transformation into a chemical composition approximating low-rank coal. IR spectroscopy performed using both DRIFTS and ATR techniques found that aromatic C-C, aromatic C-H, and carbonyl features increased with milling time, whereas alkyl and furanic features decreased. Mass spectrometry confirmed a transition from alkyl characteristics to aromatic and black carbon characteristics in parallel with an increase in carbonyl and carboxylic acid features. Raman microscopy was used to assess mid-range order associated with localized carbon structures including alkyl side chains, bridge-head carbons, and aromatic domains. X-ray diffraction confirmed that the milled hydrothermal char remained highly amorphous. Taken together, these various data sets suggest a chemical mechanism consisting of surface oxidation in parallel with radical formation due to breaking of weak aliphatic bonds followed by recondensation into stable aromatic structures. The radical formation/recondensation mechanism was then investigated using electron spin resonance, which confirmed that ball milling initially increases radical concentrations by about a factor of 10 in the first 30-60 min followed by a gradual decrease over the next several hours. In future work, we will use the fundamental information gathered in this study to design processes for functionalizing hydrothermal char for specific applications.

Supercritical Water Hydrolysis of Biomass

Manuk Colakyan[1], Danilo Cantero[1,2] and María José Cocero Alonso[3]

(1) *Renmatix, Inc. 660 Allendale Road, King of Prussia, PA 19406*
(1) *Cornell University, Department of Chemical and Biomolecular Engineering Ithaca, NY 14853*
(2) *University of Valladolid, Department of Chemical Engineering and Environmental Technology, Valladolid, Spain*

Enablement of the bio-based industry requires access to high volume, low cost sugars, from a wide variety of biomasses. Today's bio-based chemical and fuel industries have been prevented from achieving significant market penetration, not because of gaps in the technology required, but because of economics. Traditional sugar sources, like corn and cane, are expensive feedstocks for producing relatively low value, high volume products like fuels and chemicals. The traditional second generation technologies (acid, enzymes, and solvents) that were designed to extract these low value cellulosic sugars lack the practical economics to even compete with first generation sugars, let alone traditional petrochemical sources; in fact, conventional cellulosic technologies are significantly more expensive. This reality has severely limited market adoption and integration of cellulosics. The use of supercritical water to deconstruct biomass, provides significantly cost-advantaged cellulosic sugars by using primarily water. In this presentation we will present data from a continuous pilot plant located at the University of Valladolid, Spain, capable of continuously working up to 400ºC and 30 MPa with residence times from milliseconds up to forty seconds for carrying out supercritical water hydrolysis.

The kinetics of cellulose decomposition and its dependence on temperature, pressure, and concentration and residence time will be discussed. Our findings show that at extremely low reaction times (0.02 s), cellulose is hydrolyzed in supercritical water (T=400°C and P=25 MPa) to obtain a sugar yield higher than 95 wt%, whereas the 5-hydroxymethylfurfural (5-HMF) yield is lower than 0.01 wt%. If the reaction time is increased to 1 s, the main product is glycolaldehyde (60 wt%). Independent of the reaction time, the yield of 5-HMF is always lower than 0.01 wt%.

To evaluate the reaction mechanism of biomass hydrolysis in pressurized water, several parameters (temperature, pressure, reaction time, and reaction medium) were studied for different biomasses (cellulose, glucose, fructose, and wheat bran). It was found that the H^+ and OH^- ion concentration in the reaction medium as a result of water dissociation is the determining factor in the selectivity. The reaction of glucose isomerization to fructose and the further dehydration to 5-HMF are highly dependent on the ion concentration. By an increase in the pOH/pH value, these reactions were minimized to allow control of 5-HMF production. Under these conditions, the retro-aldol condensation pathway was enhanced, instead of the isomerization/ dehydration pathway.

The use of supercritical water hydrolysis enables the depolymerization reactions to occur very fast relative to unwanted glucose decomposition as it is shown in Figure 1. Also, cellulose depolymerization rate constants using supercritical hydrolysis can be up to 7 orders of magnitude faster compared to reactions occurring at 150°C. Consequently the reactor size to carry out the hydrolysis can be reduced significantly.

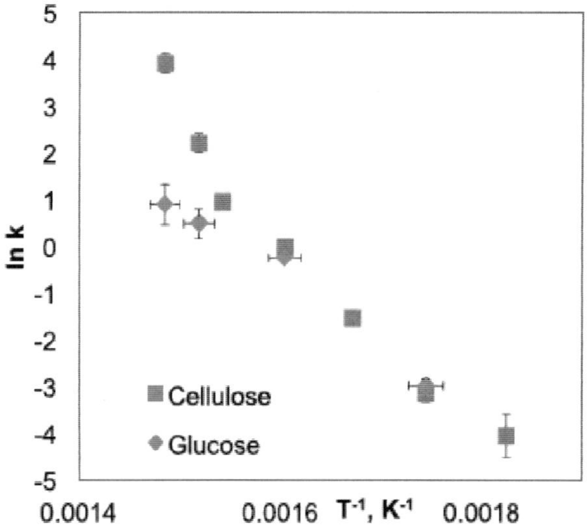

Figure 1) Kinetics constant of glucose hydrolysis in pressurized water vs. the reciprocal of temperature

The study also shows that using biomass (wheat bran) produces sugar yields > 75% with the co-product being lignin. In contrast, pure cellulose hydrolysis produces higher yields at much lower residence times. In figure 2, the evolution of sugars recovery with reaction time is displayed for what bran.

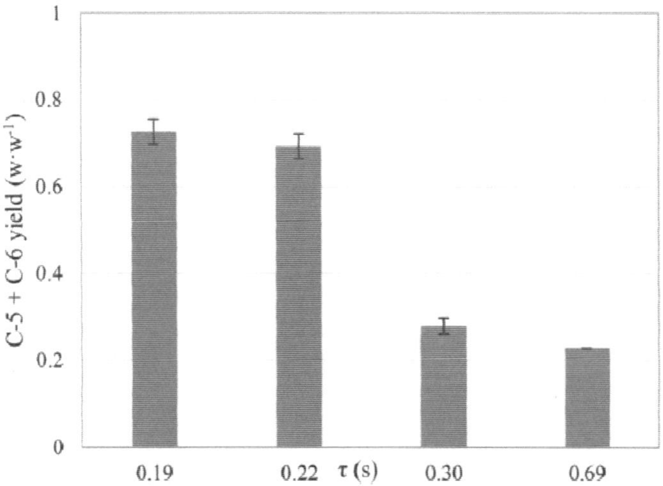

Figure 2) Yield of cellulose and hemicellulose recovered with reaction time after supercritical water hydrolysis at 400 °C and 25 MPa.

Finally, we will show the realization of this concept in a much larger scale by introducing Renmatix, Inc.'s Plantrose™ process for converting biomass to sugars. To date, no research has been reported using biomass at scales Renmatix has realized, and none have been as successful in reaching the conversion efficiencies and cost advantage that the Renmatix team has accomplished. The presentation will highlight the Plantrose™ Process, the state of the art R&D laboratories and conversion equipment at its headquarters in King of Prussia, PA, and the recently expanded Integrated Plantrose Complex (IPC) operations facility in Kennesaw, capable of converting 3 TPD (bone dry) biomass to sugars.

NREL
NATIONAL RENEWABLE ENERGY LABORATORY

Characterization of products from base catalyzed depolymerization of lignins to determine changes in chemical structure

Dr. David K. Johnson

Biosciences Center, National Renewable Energy Laboratory, Golden, CO 80401

AIChE Annual Meeting, Salt Lake City, UT

Wednesday, November 13th, 2015

NREL is a national laboratory of the U.S. Department of Energy, Office of Energy Efficiency and Renewable Energy, operated by the Alliance for Sustainable Energy, LLC.

Outline

- Introduction to lignin and lignin-derived products
- What is base catalyzed depolymerization (BCD) of lignin?
- What chemical changes occur in the lignin structure during BCD?
- Model compound studies of the BCD reaction
- Summary

NATIONAL RENEWABLE ENERGY LABORATORY 2

Lignocellulosic Biomass

Lignin: 10%–30%

- Complex aromatic structure
- Heterogeneous polymer with a variety of interunit linkages
- Resists biochemical conversion

Hemicellulose: 23%–32%

- Xylose is the second most abundant sugar in the biosphere
- Polymer of 5- and 6-carbon sugars

Cellulose: 38%–50%

- Most abundant form of carbon in biosphere
- Polymer of glucose, good biochemical feedstock

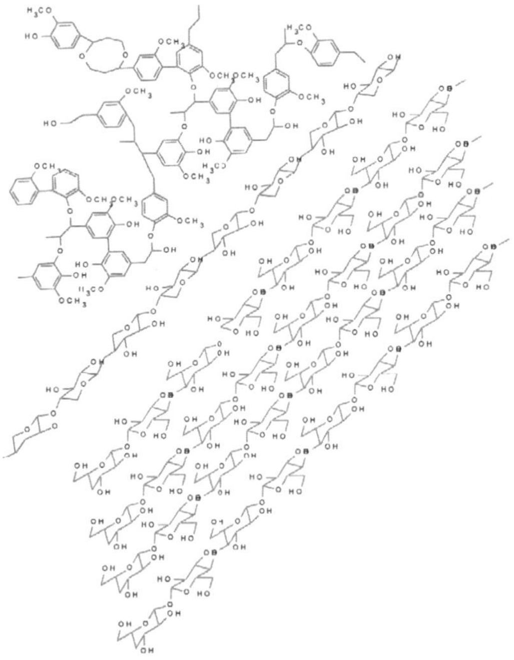

Lignin Component

- Lignin is highest in softwoods, lowest in grasses
- Lignin is the highest energy component of biomass (9,000-11,000 Btu/lb vs. 7,300-7,500 Btu/lb for cellulose)
- A $50x10^6$ gal/yr ethanol plant consuming 2000 tpd of biomass will likely have 250-500 tpd of lignin by-product (in 500 – 700 tpd residue, on a dry basis).

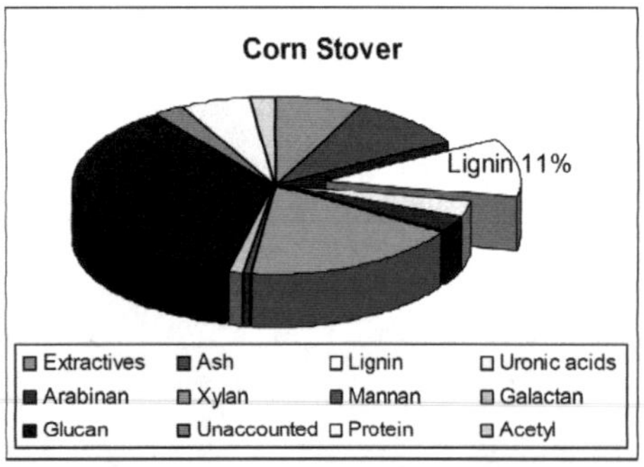

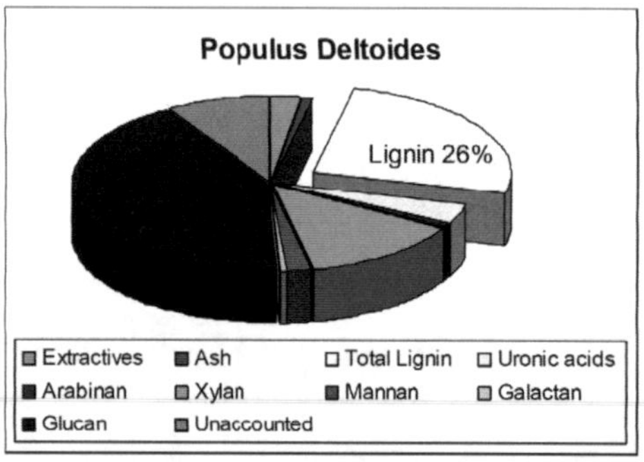

Source of Lignin in Biofuels Processes

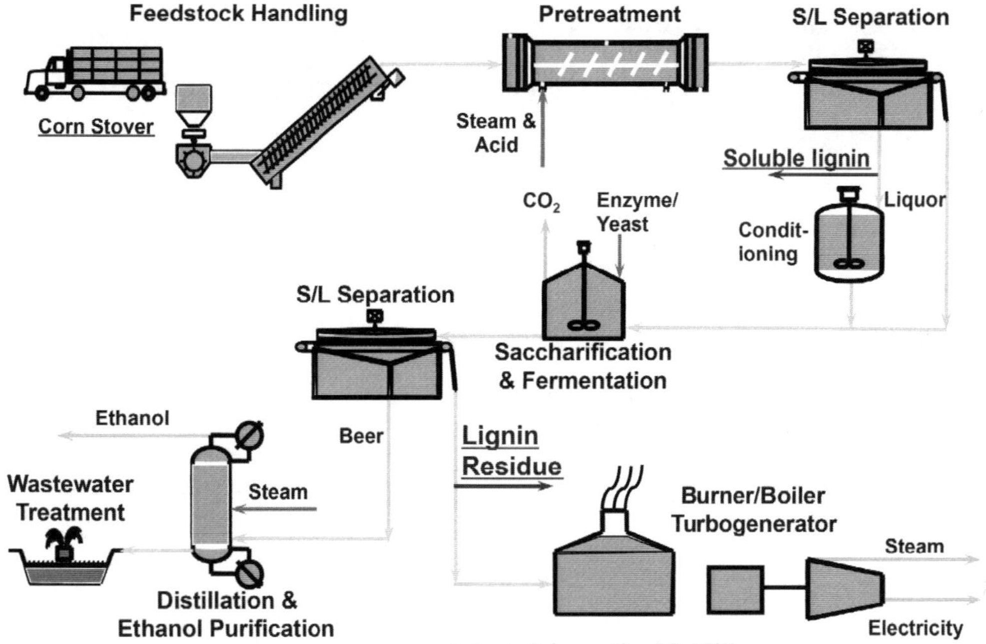

Adapted from Jim McMillan

Biorefinery Products Strategy

- A biorefinery must extract the highest value from all biomass components
- A biorefinery producing only fuels has insufficient ROI to justify construction.
- A biorefinery making only high value chemicals could generate a significant ROI, but energy impact is low.
- Strategic goals are best met by biorefineries producing both fuels and products.

Can Lignin be Transformed Into Valuable Products

"Top-Ten" Products from Lignin report (2007)

- Identifies research opportunities for lignin-derived products in an integrated biorefinery.

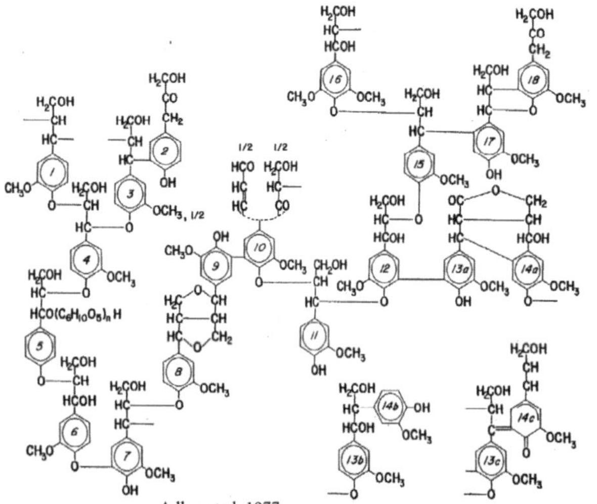

Adler, et al, 1977

Potential Lignin-derived Fuel Products

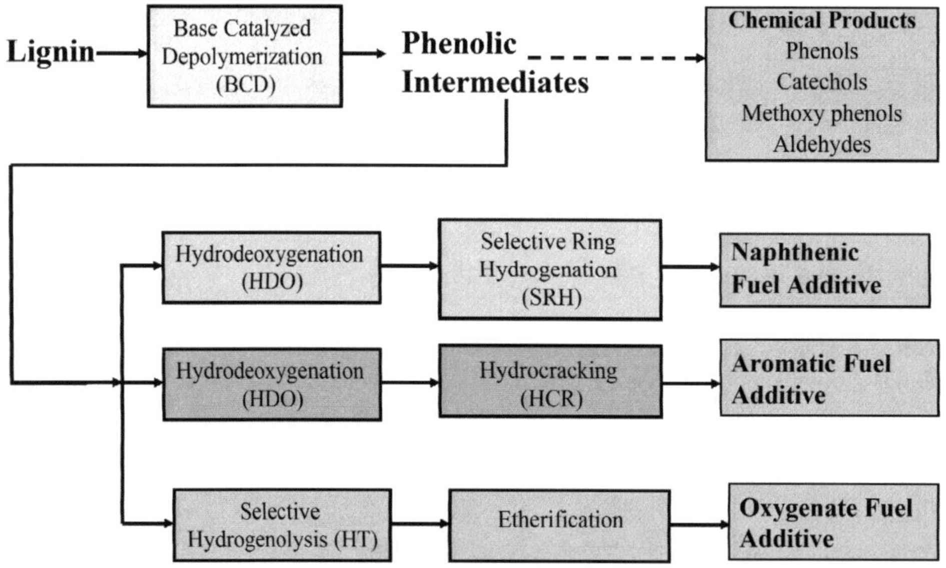

Biological Funneling Enables Selective Lignin Valorization

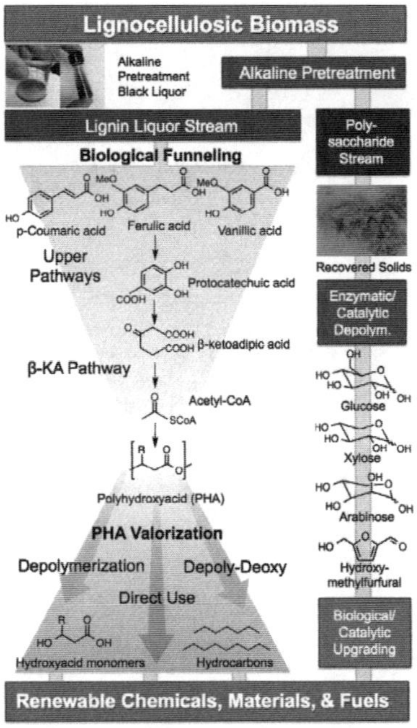

Linger, Vardon, Guarnieri, Karp, *et al.* Proc. Natl. Acad. Sci., 2014

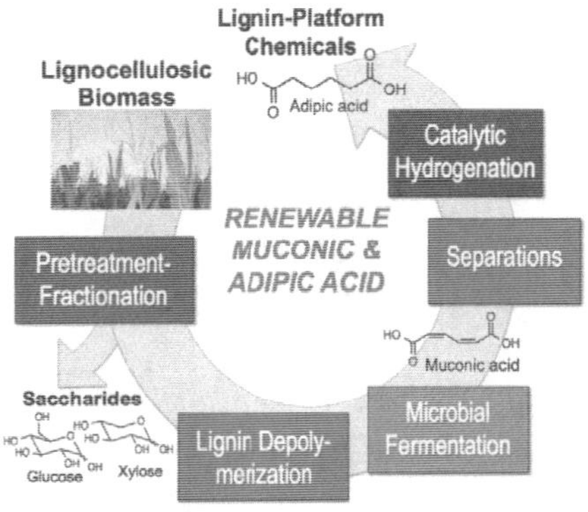

Vardon, Franden, Johnson, Karp, *et al.* Energy Env. Sci., 2015

Reaction Conditions for Aqueous BCD

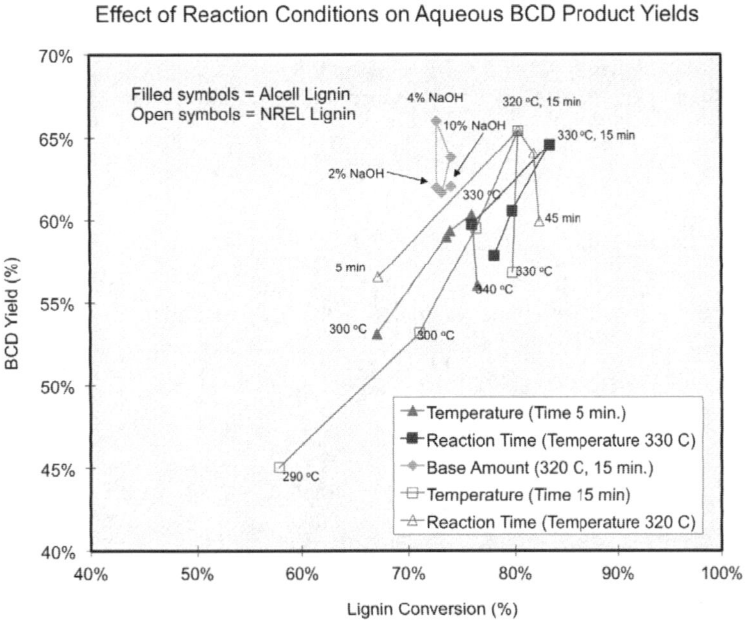

BCD Product Work-up Flowsheet

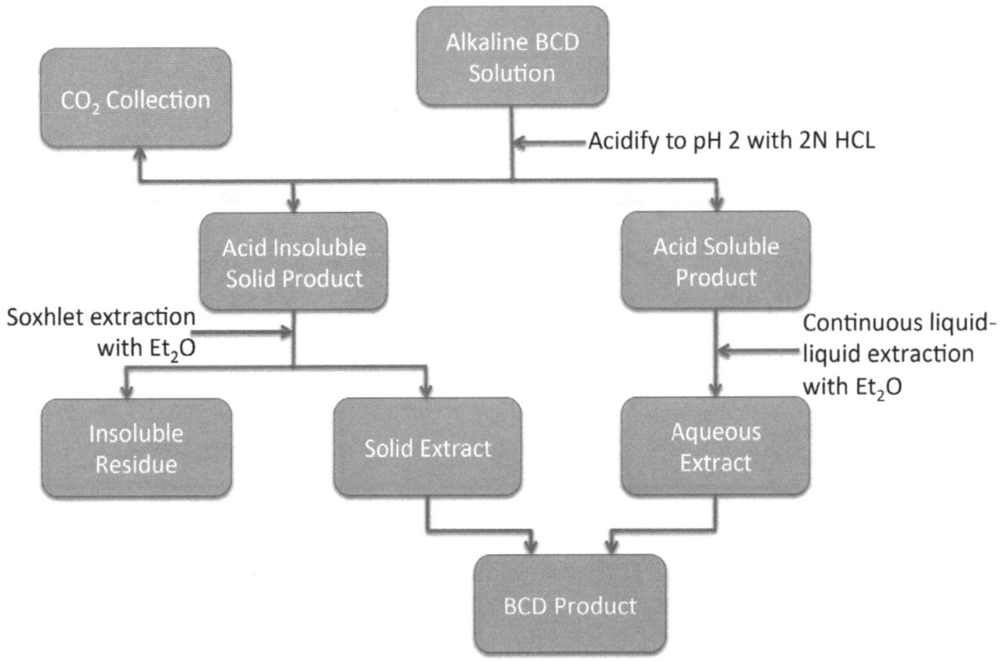

Questions to be Answered by Characterization of Depolymerized Lignin

1. How much depolymerization occurred?
   - Do different methods corroborate the change in mol. wt.?
   - What was the mechanism of depolymerization
   - What monomeric products were formed?
2. What chemical changes occurred during the reaction?
   - Was there demethoxylation or demethylation of methoxyls?
   - What happened to the lignin side-chain?

Properties of Lignin and BCD Products

	Alcell Lignin	LZD-7	LZD-19
Mn	870	470	530
Mw	2400	680	830
Volatiles (by GC)	0%	36% ± 14%	33% ± 13%
^{13}C NMR data (Number of Cs per 6 C_{Ar})			
Aliphatic C	2.9	2.6	2.5
Methyls	0.6	0.9	0.9
C=O	0.1	0.4	0.5
Methoxyls	1.3	0.7	0.3
Ar-OH	0.8	1.7	1.6
Elemental Analysis Data			
C%	66.6	68.8	70.9
H%	5.9	6.5	6.4
0% by difference	27.5	24.5	22.6

Mn = Number-average molecular weight of peracetylated material

Mw = Weight-average molecular weight of peracetylated material

^{13}C NMR Analysis of Lignin and BCD Product

- Quantitative spectrum acquisition (300 MHz spectrometer)
- Solvents d6-DMSO or d8-dioxane (0.5 – 1.0g in 3 mL)
- 10 s recycle delay time (should be 5x the longest T1)
- π / 2 pulse for maximum signal intensity, 5000 scans
- Inverse – gate decoupling (for proton decoupled spectrum)
- CH_3, CH_2 and CH subspectra generated DEPT pulse sequence

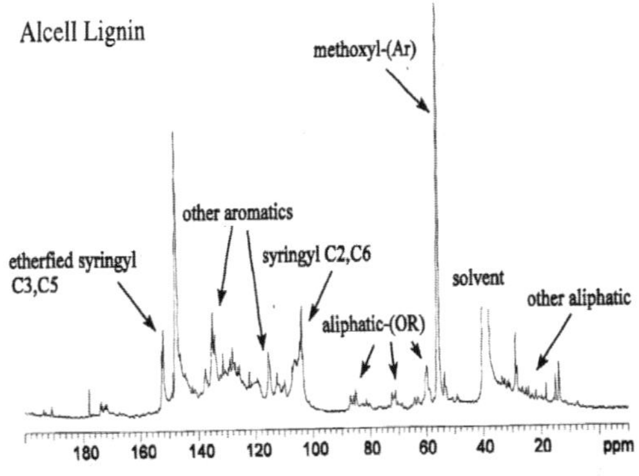

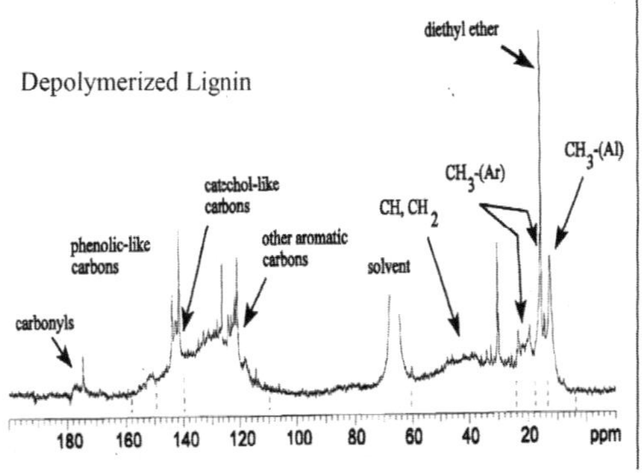

Molecular Weight Distribution by GPC

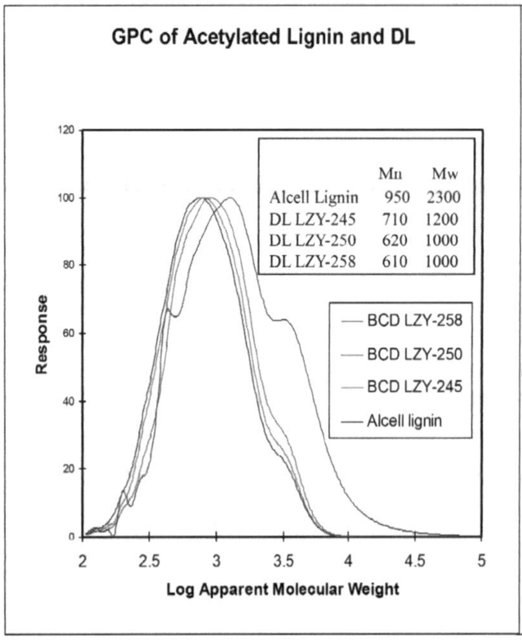

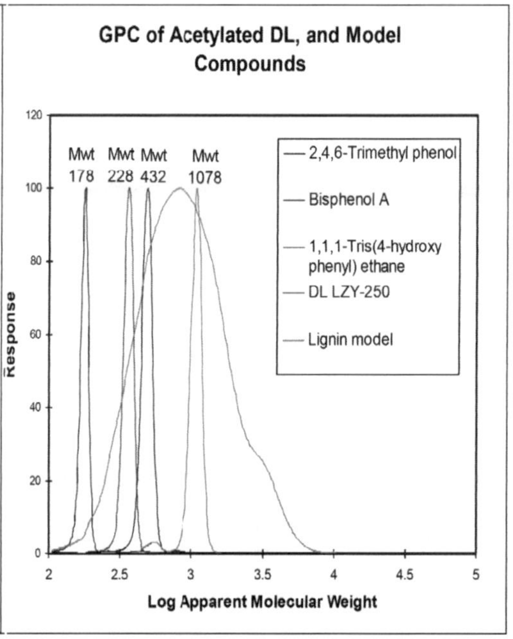

Analysis of Low Molecular Weight Material by GC

- Compounds of similar chemistry normally have similar FID response factors (RF).
- 17 phenols and catechols had avg RF of 1.9 ± 0.4
- GCs integrated from solvent peak to end of run after subtraction of a background chromatogram to allow for column bleed.
- Wt% of volatilizable material calculated from the integration area, avg RF, and sample concentration.
- Fraction volatilizable material in DL products was 30 – 35% ± 7%

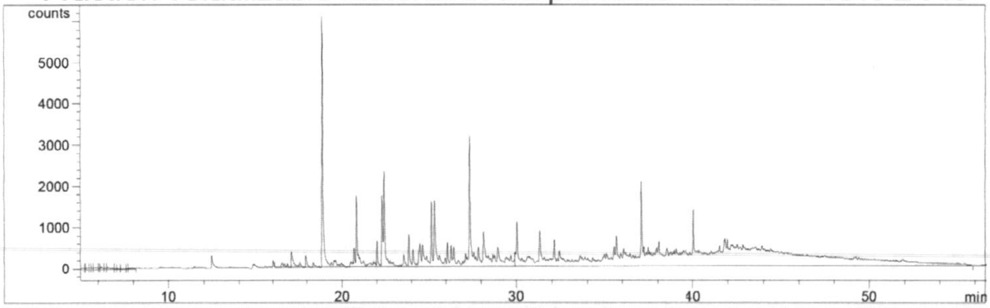

ID of Low Molecular Weight Products by RP-HPLC

- Reverse-phase HPLC used to confirm the presence of components identified by GC/MS.
- More polar components chromatograph better by HPLC than GC.
- RP-HPLC used to identify non-volatile components in the DL products using a diode-array detector and their UV spectra.
- UV spectra can be used to identify and confirm components

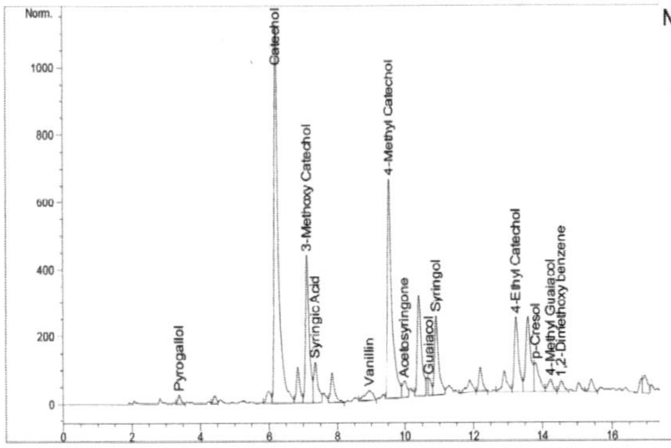

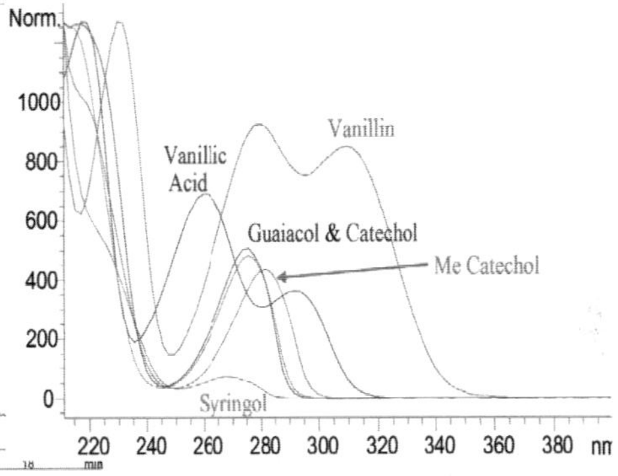

NATIONAL RENEWABLE ENERGY LABORATORY 17

Monophenolics Present in BCD Products

On acidification the lignin-derived product partitions into a solid phase and the aqueous phase

Valuable monophenolic products are mostly present in the aqueous phase

	Alcell lignin		NREL lignin	
Content (wt%)	Extract of Aqueous	Extract of Solid	Extract of Aqueous	Extract of Solid
Catechol	18.8%	0.6%	15.7%	0.7%
4-Methyl Catechol	7.3%	0.8%	6.7%	0.9%
4-Ethyl Catechol	3.4%	0.5%	2.2%	0.5%
3-Methoxy Catechol	2.4%	0.3%	2.6%	0.3%
4-Methyl Guaiacol		0.3%		0.3%
4-Ethyl Guaiacol		0.3%		0.3%
1,2-Dimethoxy benzene		0.6%		0.5%
Vanillin	0.4%	0.2%	0.7%	0.3%
Syringol / Guaiacol	2.0%	1.2%	1.3%	1.1%
4-Methyl Syringol		0.4%	0.4%	0.3%
Syringealdehyde		0.2%	0.4%	0.4%
Acetosyringone	0.9%	0.6%		0.5%
Total	35.2%	6.1%	30.0%	6.1%

NATIONAL RENEWABLE ENERGY LABORATORY 18

Answers Provided by BCD Characterization

- Was there demethoxylation or demethylation of methoxyls?
  Evidence is clear in the NMR data for mostly demethylation
- What happened to the lignin side-chain?
  NMR indicates loss of O from the side-chain
  Volatile fraction did not contain products with O side-chains
- How much depolymerization occurred?
  GPC and GC indicate that lignin was partially depolymerized.
  Monomeric phenolics were made but oligomers remained

Average Reaction Chemistry for BCD Reactions

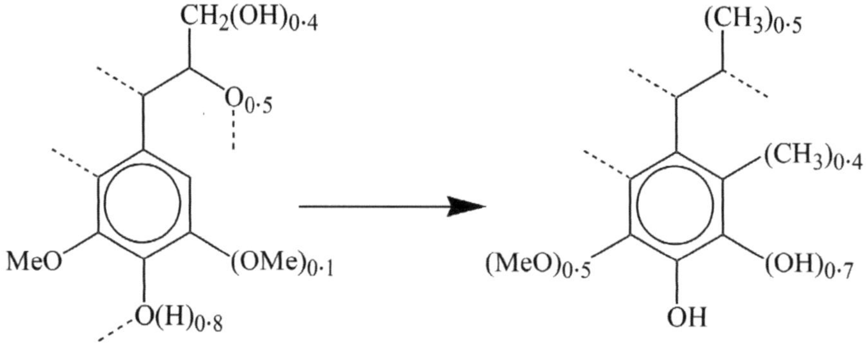

Alcell Lignin	Depolymerized Lignin
66.8% C 5.9% H 27.0% O	70% C 6.5% H 23.5% O
$C_{8.9}H_{6.3}O_{0.7}(CH_3)_{0.6}(OH_{Ph})_{0.8}$ $(OH)_{0.4}(OMe)_{1.1}(CO)_{0.1}$	$C_{8.2}H_{4.8}(CH_3)_{0.9}(OH_{Ph})_{1.7}$ $(OMe)_{0.5}(CO)_{0.4}$
Repeat Unit MWt. = 182,	Repeat Unit MWt. = 170,
Oligomer range 2-mer to 9-mer	Oligomer range 1-mer to 3-mer

Study of Reactivity of Methoxy Phenols to BCD

Questions to be answered by Model Compound Reactions

- How stable are potential BCD products to reaction conditions
- Are oligomers formed
- $NaOH \longrightarrow Na_2CO_3$ where does the CO_2 come from.

Model Compound Reaction Conditions

- Tubing (150 mL) reactor, heated in fluidized sand bath (at 295 or 330°C) for 1 or 10 min.
- Aqueous NaOH solutions (1% and 3% by weight)

Does oxidation occur during alkaline hydrolysis.

Are side-chain C & O lost

Does CO_2 come from the side-chain

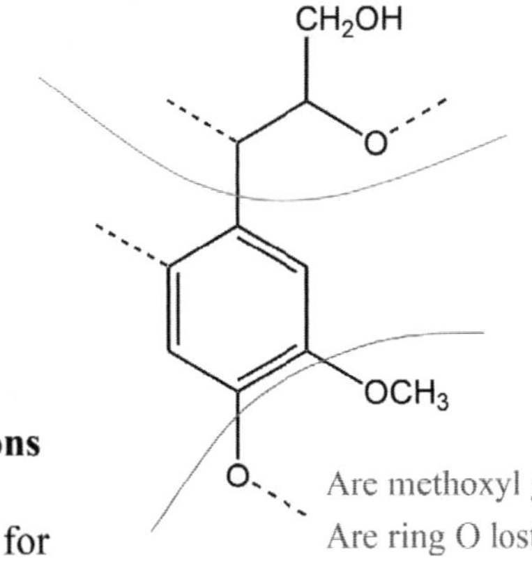

Are methoxyl groups lost

Are ring O lost

Reactions of Guaiacols

- Conversions in 20 – 90% range (severity had greatest effect)
- Main reaction was demethylation giving catechols
- Dimethoxy benzenes are also formed
- Transmethylation observed
- Little evidence of CO_2 formation
- Low levels (5 – 20%) of oligomers formed

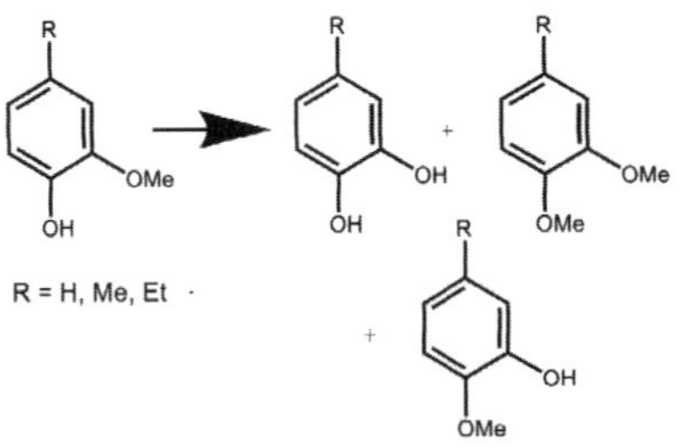

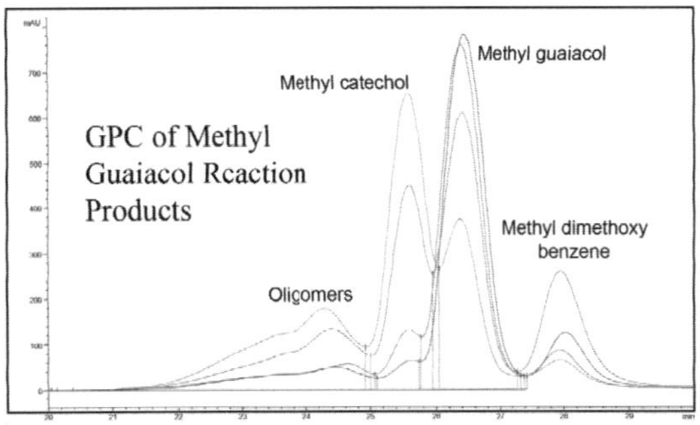

Reactions of Vanillic Acids

- Conversions 100% except for homovanillic acid at lowest severity
- Main reaction is decarboxylation giving guaiacols then catechols
- Quantitative formation of CO_2 with acidification of product soln.
- Dimethoxy benzenes are also formed
- Transmethylation observed
- Low levels (0 – 15%) of oligomers formed

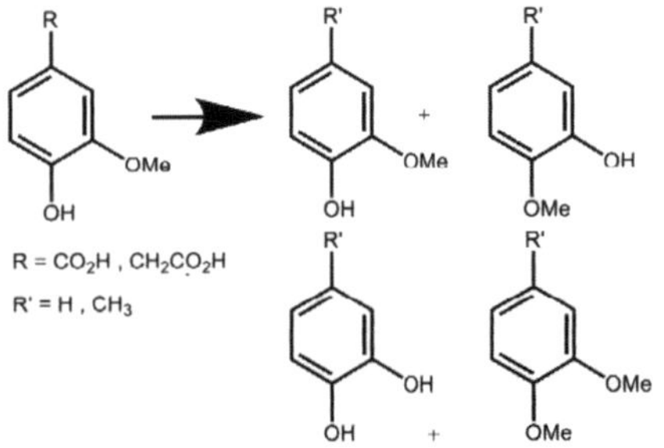

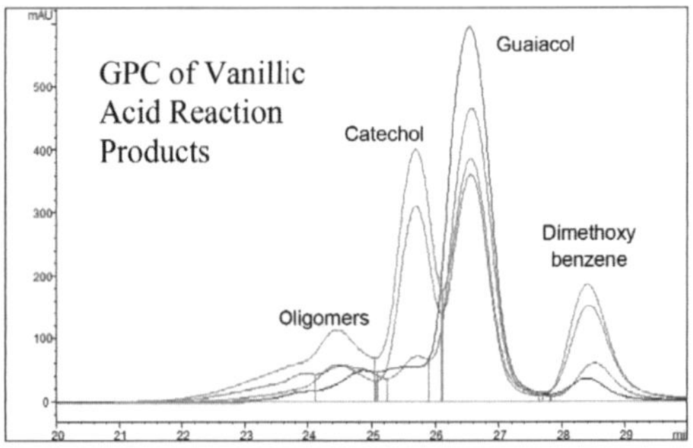

NATIONAL RENEWABLE ENERGY LABORATORY 23

Reactions of Vanillin

- Conversions <40% at lowest severity, >90% at highest severity
- Decarbonylation, demethoxylation, and oligomer formation all observed
- Significant formation of CO_2 (15 mole%) and oligomers (26%) mostly seen at low base/high severity condition (green curve)
- Products observed that still contain carbonyl group
- No oxidation products observed

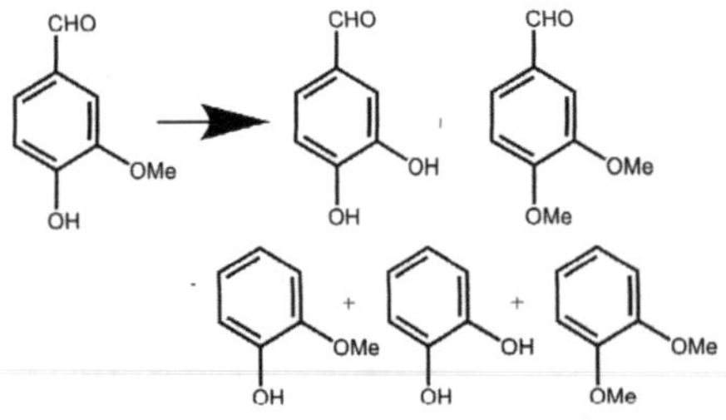

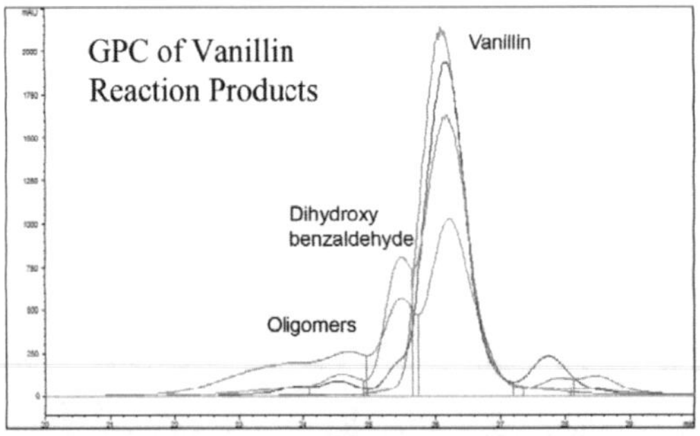

NATIONAL RENEWABLE ENERGY LABORATORY 24

Reactions of Catechol

- Conversion increased with base concentration and severity
- Main reaction is formation of oligomers, no formation CO_2 observed
- Oligomers have much smaller response in the RI detector than the UV detector suggesting they contain more chromophoric groups
- GPC of acetylated products indicates that the average molecular weight of the oligomers (390) is twice that of catechol (194).

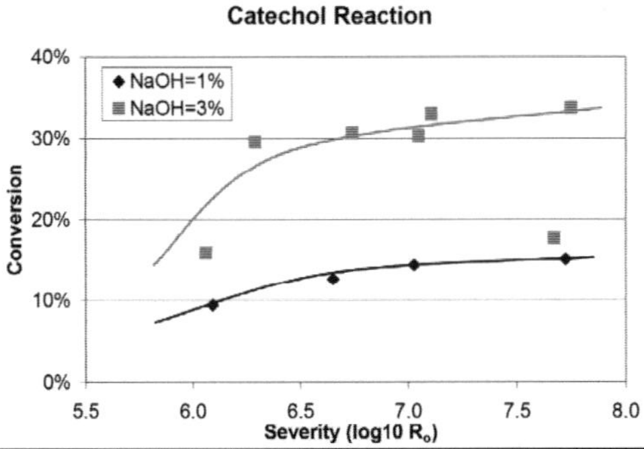

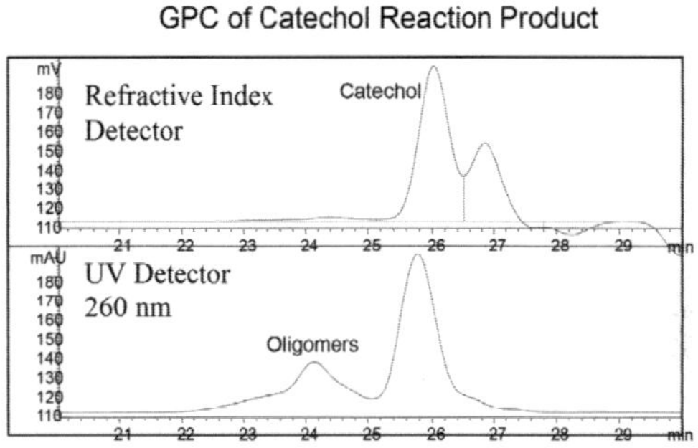

Reactions of Ether-linked Dimers

Conversions at 330°C, 15 min	1.4% NaOH	0.05% NaOH
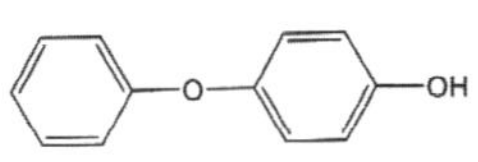	41% (16% phenol)	5% (3% phenol)
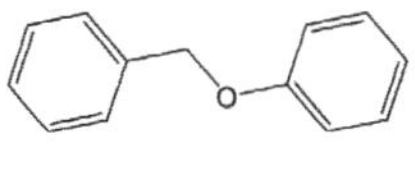	74% (48% phenol)	61% (20% phenol)
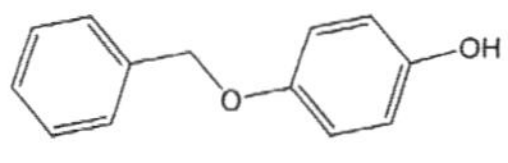	100%	100%
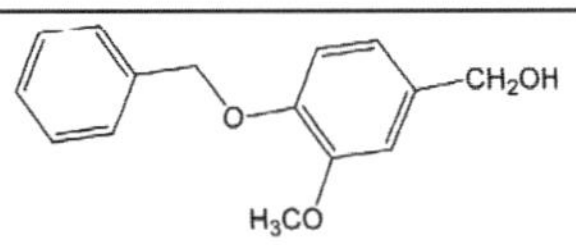	100%	100%

- Conversion of benzyl ethers easier than phenyl ethers
- Oxygen on the ring increases conversion
- Ether cleavage even at low NaOH concentrations

Reactions of Lignin Model Dimers

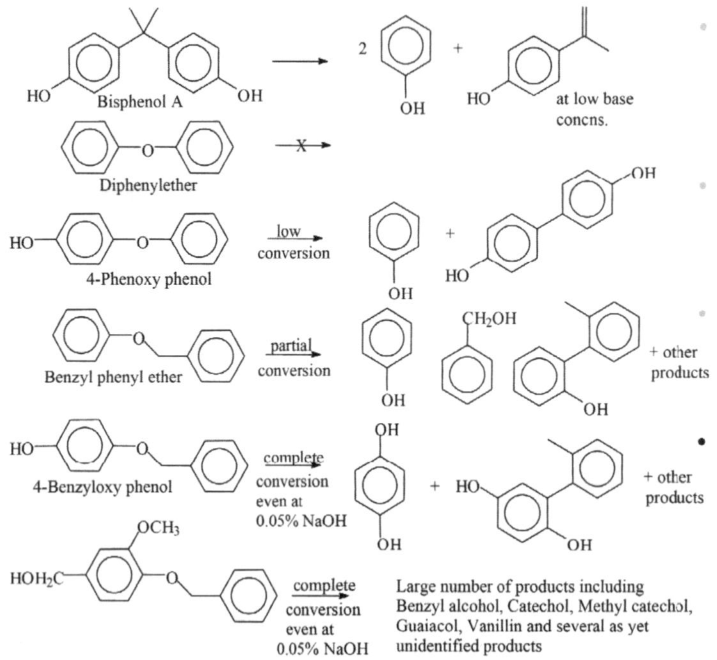

- Conversion of benzyl ethers easier than phenyl ethers
- Oxygen on the ring increases conversion
- Ether cleavage even at low NaOH concentrations
- Some biphenyl products formed

Summary

- Benzyl ether linked dimers can be cleaved with 0.05% NaOH. Diphenyl ether dimers are more resistant. Some biphenyl products observed.
- Extent of depolymerization appears to be limited by some condensation reactions involving C-C bond formation

Oxidation occurs during BCD
Side-chain is C & O lost
CO_2 comes from the side-chain probably decarboxylation of carboxylic acids

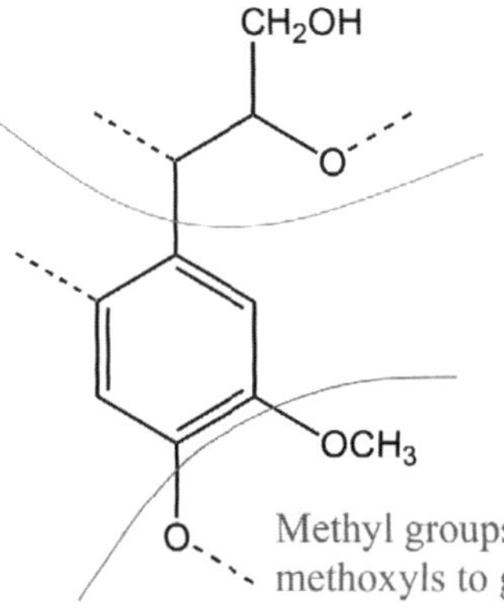

Methyl groups lost from methoxyls to give catechols
Little evidence for loss of ring oxygens

Acknowledgements

NREL
Esteban Chornet
Justin Sluiter
Mark Davis
Joan Salvado (on sabbatical from U. Tarragona, Spain)

University of Utah
Professor Joseph Shabtai, Dr Wlodek Zmierczak

US DOE Bioenergy Technology Office (BETO) for funding this work.

NREL National Renewable Energy Laboratory

437220 Hydrolysis of Ddg

Abstract:

Liquid hot water (LHW) extraction was used as a pretreatment method to separate the hemicellulose fraction from dried distiller's grain with solubles (DDGS) into liquid phase. Acid hydrolysis using 3.264% w/w sulfuric acid at 130 C was performed to convert polysaccharides in LHW extract to monosaccharides. The structure characterization of DDGS in anomeric carbon region based on proton NMR and heteronuclear single quantum coherence (HSQC) during acid hydrolysis was studied in this work. It reveals that the sugar units in DDGS hemicelluloses are constructed with (1-4)-β-D-xylopyranose and α-L-arabinofuranosyl residues. A kinetic model is included to explain the changing concentration of monomer, oligomer and sugar units. The model was further tested based on the changing concentration of five carbon sugar units during hydrolysis.

437333 How Do Hemicellulosesaffect the Lignocellulose Enzymatic Digestibility

Abstract:

Hemicelluloses have been considered as a major factor limiting the enzymatic digestibility of lignocellulosic cellulose for glucose releasing. To study how hemicelluloses affect the enzymatic hydrolysis of lignocellulose, dilute acid was applied for removing hemicelluloses of wheat straw under mild conditions to obtain samples with different hemicelluloses content. To investigate the interaction of lignin with hemicelluloses, sodium chlorite was used to selectively remove lignin. It was found that with a high lignin content (28~38%), cellulose conversion increased with decrease of hemicelluloses (noted as xylan) content, but the highest cellulose conversion (~2% xylan content) was just about 65%. However, when lignin content was low (~4%), hemicelluloses showed nosignificant negative effect on cellulose hydrolysis, and the final cellulose conversion could be as high as 90% even if the xylan content was as high as 28.25%. It seemed that lignin is more negative for the enzymatic hydrolysis of lignocellulose. Xylan was added to hydrolysis of pure cellulose to study how hemicellulose affect the lignocellulose enzymatic digestibility. The results indicated that xylan, especially soluble xylan had anegative effecton cellulose hydrolysis. And the existence of xylan decreased cellulase adsorption on cellulose. Further study is needed to investigate the mechanism.

429513 High-Yield Production of Furans from Lignocellulosic Biomass Under Mild Conditions in a Biphasic Process with Molten Salt Hydrate

Abstract:

This presentation is to report an innovative biphasic process for converting lignocellulosic biomass to furans (furfural from pentoses and bromomethyl furfural (BMF) from hexoses) with high yield under mild conditions (100-130 °C and 1-3 h). The biphasic system consists of an aqueous phase of molten salt hydrate, in which cellulose and hemicelluloses in the biomass are converted to BMF and furfural, respectively, without pretreatment and an organic phase of extraction solvent, into which the furans formed in the aqueous phase are immediately extracted to avoid undesired further decomposition of the products or condensation to humins. Meanwhile, lignin in the biomass is depolymerized and separated as a solid residue from the system, which is potentially a good feedstock for lignin coproducts. The process is robust to handle different types of biomass, including corn stover, switchgrass, aspen, poplar and Douglas fir. Under the optimal conditions investigated, the molar yields of furfural and BMF were ~70% and ~85%, respectively.

437365 Study on Co-Pyrolysis Characteristics of Rice Husk and Ammonium Salts

Abstract:

The current study presents the co-pyrolysis characteristics of rice husk and different kinds of ammonium salts in a fixed bed reactor respectively. Urea, ammonium acetate, ammonium sulfate and ammonium dihydrogen phosphate were selected as ammonia source. The results indicated that the impregnation of urea and ammonium acetate could improve the yield of liquid products and deduce the yield of solid, while the impregnation of the other two kinds of ammonium salts antipodal effects. The introduction of ammonium salts reduced the content of carbonyl compounds and produced some new nitrogenous heterocyclic compounds with high value and stability in the liquid products. When ammonium sulfate and ammonium dihydrogen phosphate were impregnated, the liquid products had relatively few types of compounds and mainly contained levoglucosan and furfural. Meanwhile, the effects of ammonium salts on the gas and solid products were investigated.

434038 Progress on Furfural Production Using Xylose-Rich Hydrolysate from Hardwood Autohydrolysis Pretreatment

Abstract:

2015 AIChE Annual Meeting

International Congress on Energy (ICE)

T4A03 Chemical Conversion Processes in Forest/Plant Biorefineries

November 8-13, 2015

Salt Lake City, UT

Progress on furfural production using xylose-rich hydrolysate from hardwood autohydrolysis pretreatment

Lu Liu*, Hou-min Chang, Hasan Jameel, Sunkyu Park

Department of Forest Biomaterials, North Carolina State University, Raleigh, NC 27695, USA

* Presenting author: llui@ncsu.edu

Furfural, the dehydration product of xylose, can serve as a sustainable substitute for petroleum-based building blocks for fine chemicals and plastics. To produce furfural from biorefinery system, our study focuses on establishing an efficient and environmentally friendly method using autohydrolysis hydrolyzate from biomass.

A batch process with a vapor-releasing reactor, where furfural can be released immediately after the reaction, has been used. It was observed that by using 2% xylose solution without catalyst, over 99% on xylose conversion and approximately 65% on furfural yield were achieved. When using autolydrolysis hydrolyzate that contains more than 80% of xylose-oligomer, a final furfural yield of over 70% was observed. It is considered that the by-products from autohydrolysis such as acetic acid act as catalyst in xylose dehydration process, and the pre-existed furfural in hydrolyzate makes it an even better source for furfural production comparing to pure xylose solution.

However, in a batch process, furfural generation rate decreased exponentially along with the reaction time, which makes the conversion inefficient. A lab-scale distillation showed that the furfural recovery yield is unsatisfactory when xylose concentration in the feed is low, which makes the whole process less attractive. In order to improve furfural production, a continuous reactor with constant and high concentration xylose solution as feed was used. The optimized condition for furfural production in a continuous reactor and the possibility of directly using xylose-rich hydrolyzate will also be discussed in details.

Effect of Fast Pyrolysis Conditions on the Biomass Solid Residues at High Temperatures (1000-1400°C)

Anna Trubetskaya[1]
Angel David Garcia Llamas[2]
Associate Professor Kentaro Umeki[2]
Senior Researcher Peter Arendt Jensen[1]
Professor Anker Degn Jensen[1]
Professor Peter Glarborg[1]

[1] *DTU Chemical Engineering, Green research center, 2800 Lyngby, Denmark*
[2] *Department of Engineering Science and Mathematics, LTU, 85478, Luleå, Sweden*

DTU Chemical Engineering
Department of Chemical and Biochemical Engineering

Objectives

- Woody (pine, beech) and herbaceous biomass (alfalfa straw, wheat straw, rice husk)
- Pyrolysis of smaller and larger particle size (> 0.5 mm)
- Experimental investigations and modeling of char yield at fast heating rates (10^2-10^4°C/s) and at high temperatures (up to 1500°C)
- Potassium and silicon bearing compounds influence on the char yield, reactivity and morphology

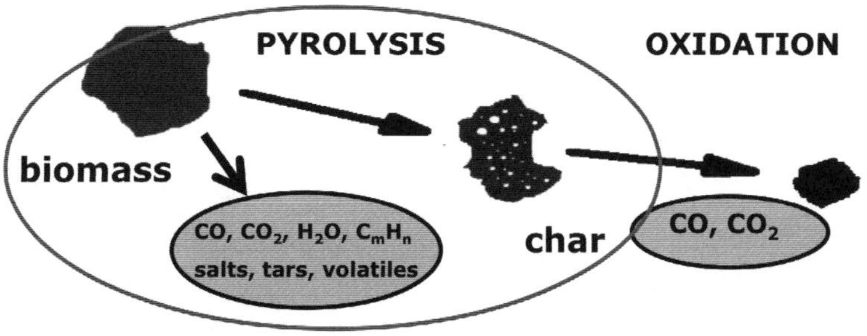

Experimental setup

Wire mesh reactor

Single particle reactor

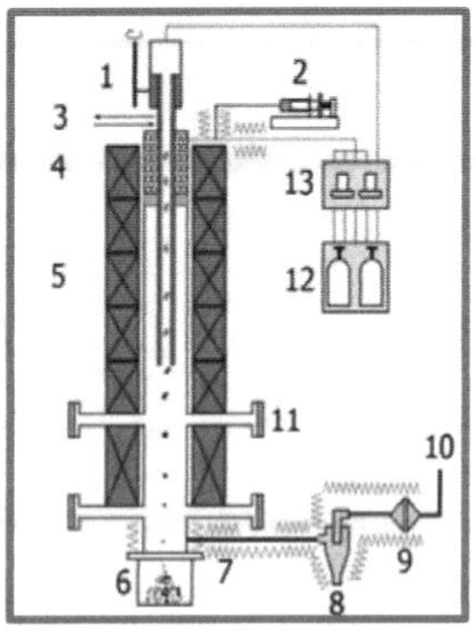

Drop tube reactor

Operational parameter	Wire mesh reactor	Single particle reactor	Drop tube reactor
Max. Temperature [°C]	1650	1500	1500
Heating rate [°C/s]	≤ 5000	≤ 200	≤ ≈ 10^4
Particle size range [mm]	≤0.65	≥ 3	≤ 1

Luleå University Drop-Tube Furnace

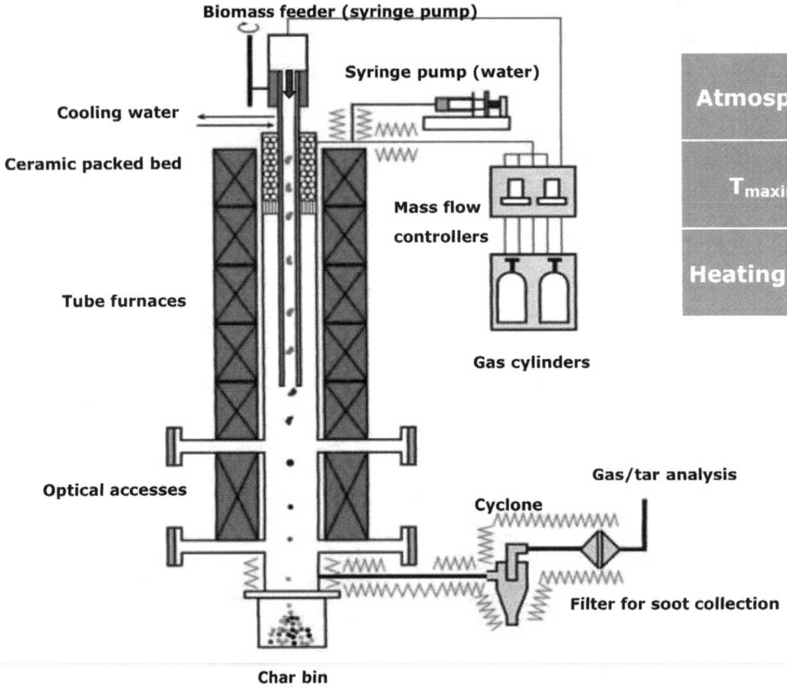

Atmosphere	N_2, O_2, H_2, CO_2, H_2O, Ar, CO
$T_{maximal}$	1500°C
Heating rates	up to 10^4°C/s

Char yield comparison

DTU

WMR (1000°C, 1 s holding time, 0.2mm) DTF (≈10^4°C/s) and WMR (10^3°C/s, 1 s, 0.2mm)

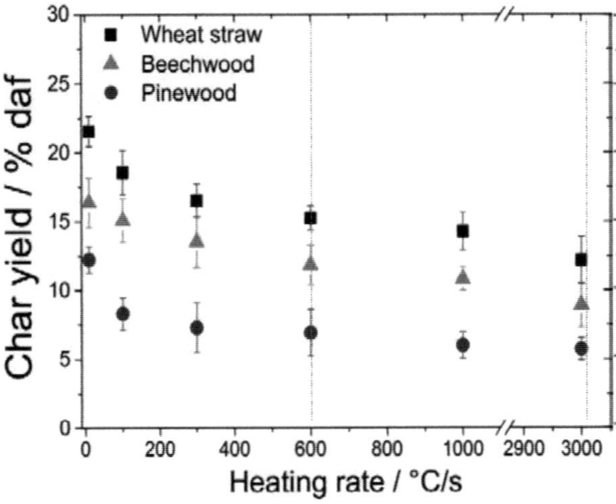

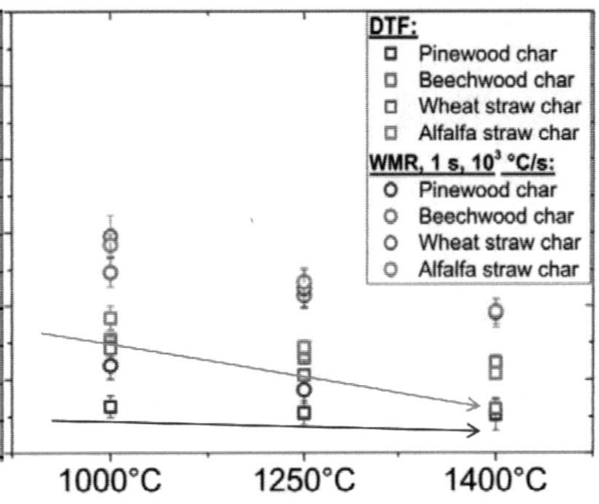

- (Graph 1) Char yields of wood and herbaceous biomass in the WMR decreased with the increasing temperature
- (Graph 1) At heating rates (> 600°C/s), the char yield is nearly constant, except wheat straw
- (Graph 2) DTF heating rates led to the char yield decrease
- (Graph 2) At final temperatures (> 1000°C) in the DTF, only wheat straw char showed 3.5% points decrease

Alkali effect on char yield

DTU

WMR, 1000°C/s, 1 s holding time DTF, ≈10^4°C/s, 1 s residence time

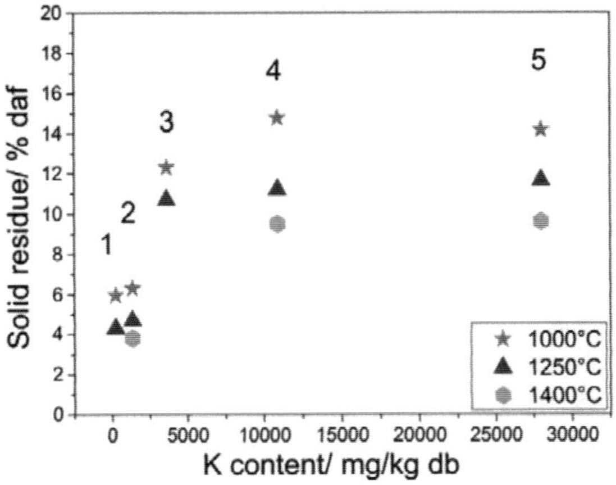

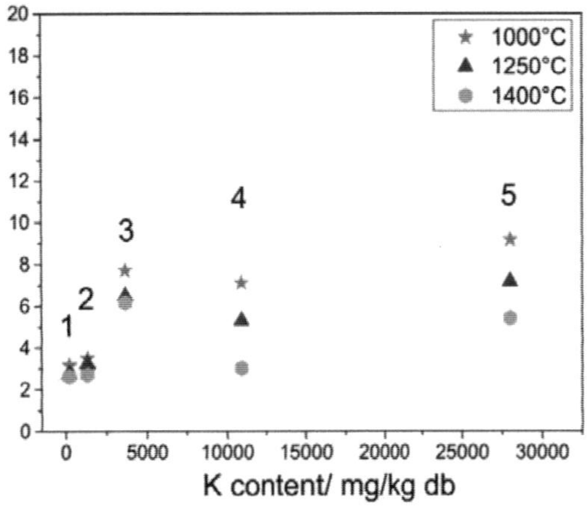

- Graphes 1 and 2 show char yield (daf) of pinewood (1), leached wheat straw (2), beechwood (3), wheat straw (4) and alfalfa straw (5) over the potassium content in original biomass
- Potassium compared to all other ash elements in the fuels had the highest influence on the char yield (daf)
- At intermediate heating rates (WMR), potassium influenced the char yield significantly more than at high heating rates in the DTF

Silicon oxides effect on char yield and reactivity

Char yield, WMR, 1000°C/s, 1 s holding time

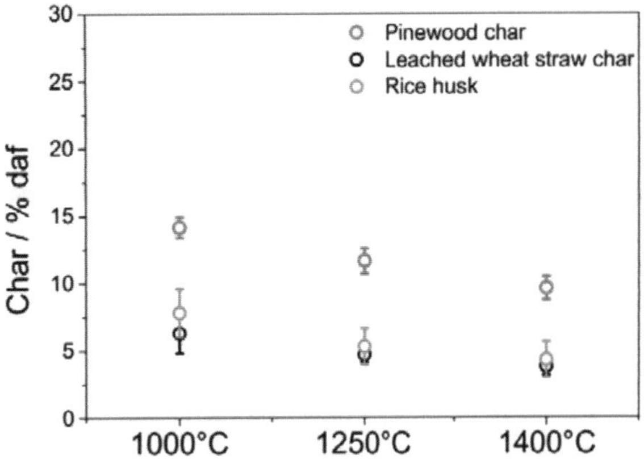

TGA, 5 vol.% O_2 + 95 vol. % N_2

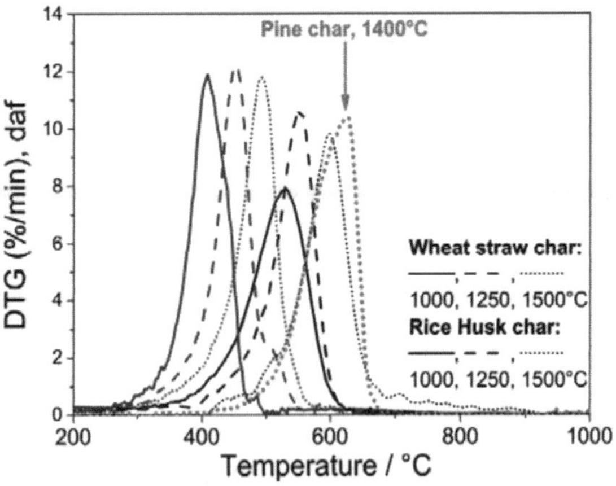

- Graph 1 shows that silica has no influence on the char yield compared to potassium
- Graph 2 shows that the reactivities of pinewood and rice husk chars were similar in oxidation, indicating less influence of silica on the char reactivity
- Graph 2 shows that the alkali rich wheat straw chars were 6, 18 and 50 times more reactive than wood and rice husk chars

Large particle effect on the char yield

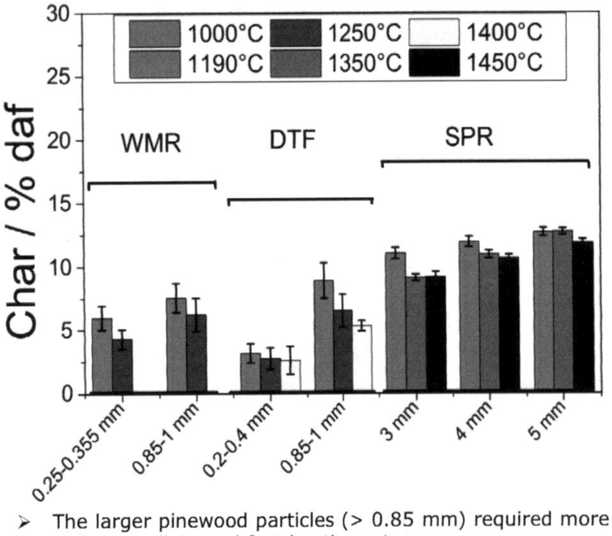

WMR:

Heating rate: 1000°C/s

Holding time: 1 s

Particle size: 0.25-0.355 mm, 0.85-1 mm

DTF:

Heating rate: 10^4°C/s

Residence time:1 s

Particle size: 0.2-0.4 mm, 0.85-1 mm

SPR:

Heating rate: 200°C/s

Particle size: 3-5 mm pine cubes

- The larger pinewood particles (> 0.85 mm) required more than 1 s holding time for the complete conversion at intermediate and fast heating rates
- The influence of heating rate on the char yields was less pronounced for particle sizes from 0.85 to 4 mm obtained at temperatures > 1000°C/s

Char and soot yields in the drop tube reactor

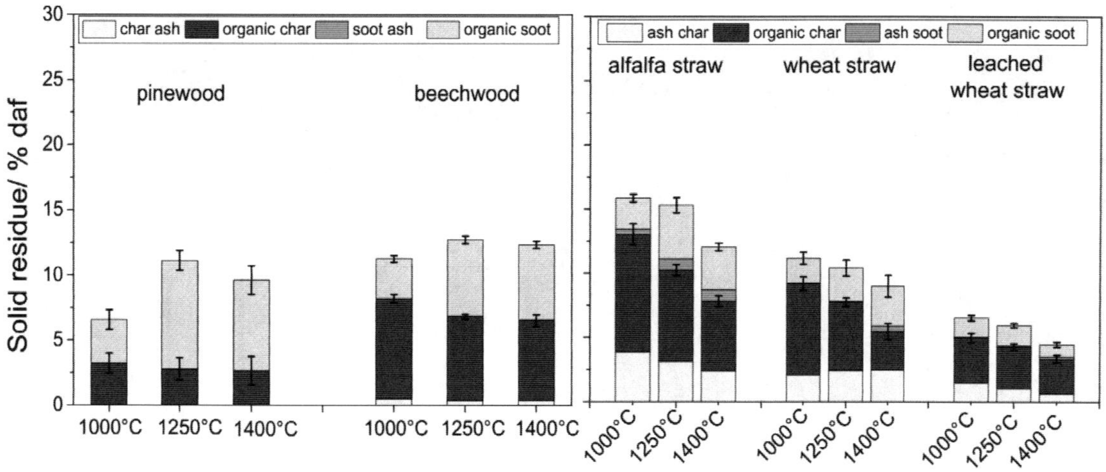

- Both graphes compared char and soot yields of woody (Graph 1) and herbaceous biomass (Graph 2)
- Lignin content of leached wheat straw decreased from 25.6 to 15.6 wt.%
- Lignin has a stronger influence on the soot formation than potassium

Soot morphology with TEM microscopy

Pinewood soot, 1250°C

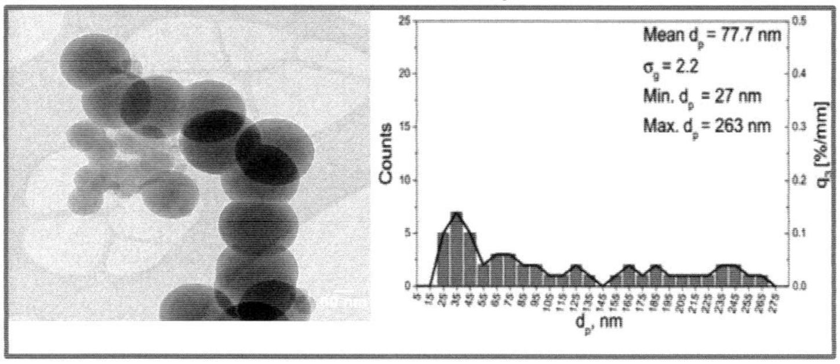

Pinewood soot, 1400°C

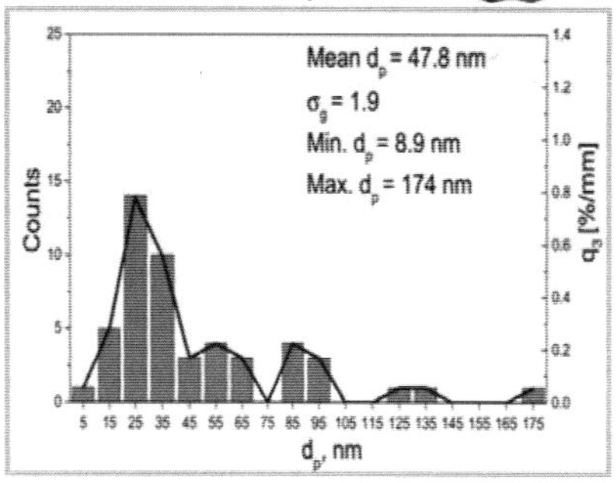

Wheat straw soot, 1250°C

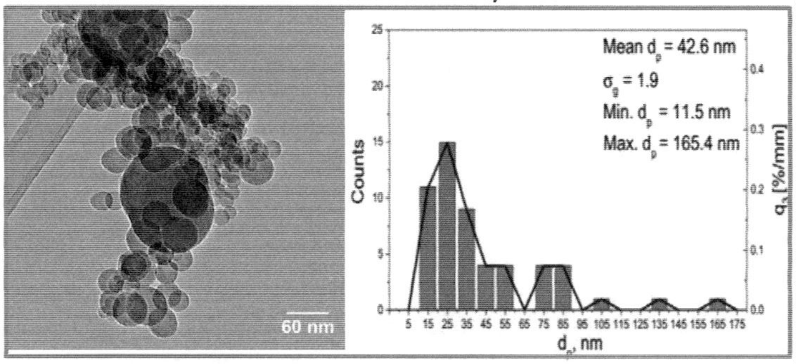

Wheat straw soot, 1400°C

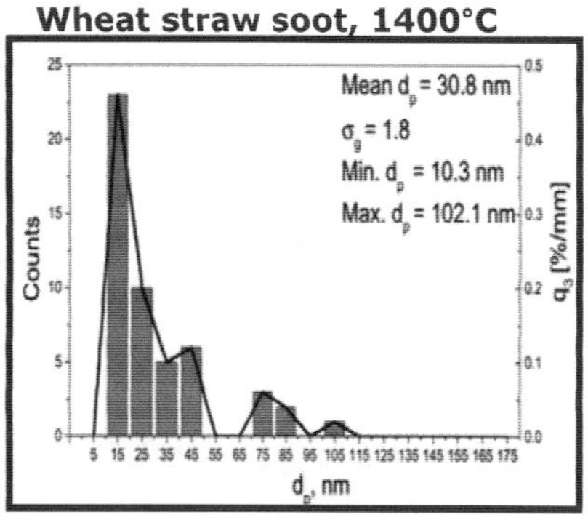

Soot morphology with TEM microscopy

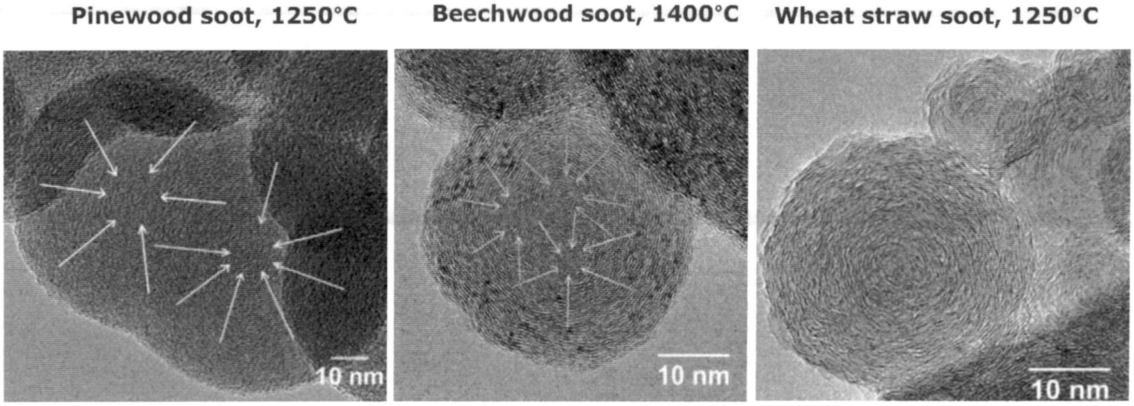

- Pinewood formed particles with the pre-dominating multi cores and located on a larger distance
- Beechwood formed a mixture of multi cores and single cores at 1250 and 1400°C
- Wheat straw soot particles were mostly with single cores
- TEM microcropy showed a more graphitic structure of beechwood and wheat straw soot particles

1D modeling of fast pyrolysis

Broido-Shafizadeh schema*

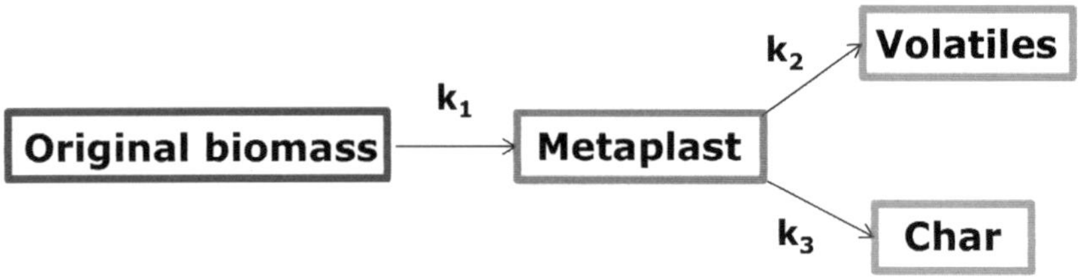

Reaction equations*:

$r_i = k_i \cdot m_i$

$k_i = -k_{0,i} \cdot exp\left(-\frac{E_{a,i}}{R \cdot T}\right)$

$E_{a,3} = E_{a,3}(\omega) = E_a \cdot (1 - (1 - C_1)) \cdot \left(1 - \exp\left(\frac{\omega}{C_2}\right)\right)^2$

$C_1, C_2 = constants$

$\omega = potassium\ content$

Initial boundary conditions*:

$m_{BM}(0) = 1$

$m_{MP}(0) = m_{VM}(0) = m_{Char}(0) = 0$

$T_{particle}(0, r) = T_{amb}$

*Broido et al., 1971, Bradbury et al. 1979 et al.

1D modeling of fast pyrolysis

DTU

Pinewood 0.2mm particles

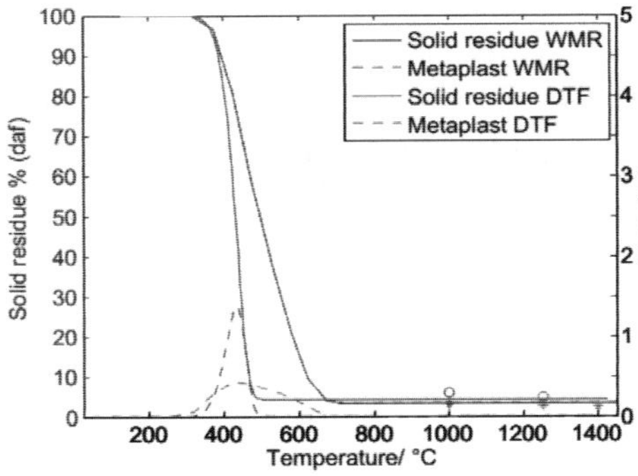

Wheat straw 0.2 mm particles

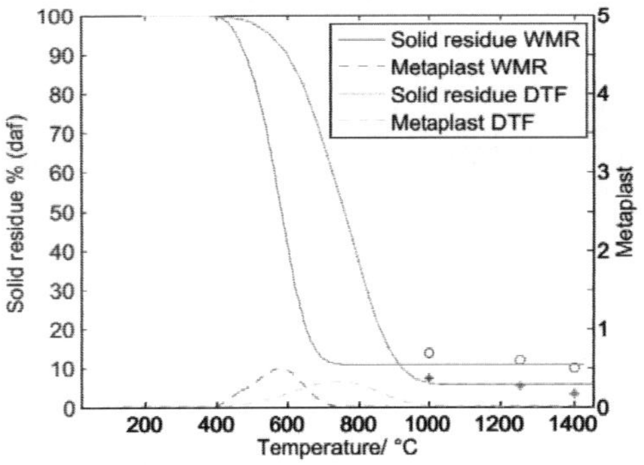

Advantages:

- Internal thermal gradient
- Potassium effect on the char yield
- A simple 1D model
- Unique kinetic parameters

Improvement:

- No differences in the char yield of smaller and larger particles => secondary reactions
- Oversimplifications

Pinewood 1mm particles

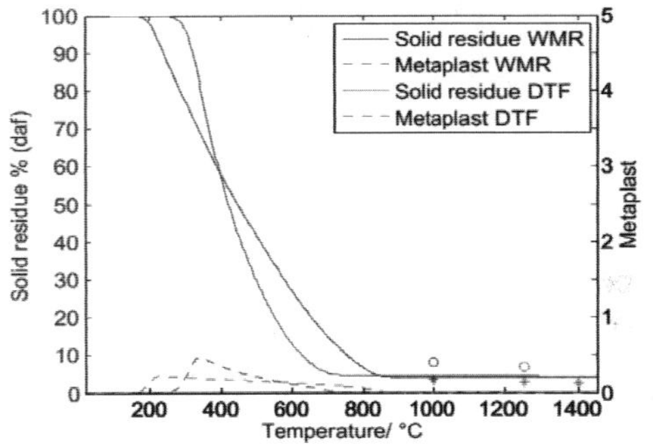

Summary

DTU

- The heat treatment temperature and potassium content affected the char yield stronger than the heating rates and differences in the plant cell wall compounds between 600 and 3000°C/s
- Potassium compared to all other ash elements in the fuels had the highest influence on the char yield
- At intermediate heating rates (WMR), the catalytic effect of potassium was more pronounced than at high DTF heating rates
- Low lignin content leads to the lower soot yields
- The proposed kinetic model for the fast biomass pyrolysis is relatively simple and predicts reasonably the char yield of wood and herbaceous biomass particles < 10 mm

Acknowledgements

- Department of Engineering Science and Mathematics, LTU
- Dong Energy and Vattenfall

DONG energy

VATTENFALL

LULEÅ UNIVERSITY OF TECHNOLOGY

DTU Chemical Engineering
Department of Chemical and Biochemical Engineering

Thanks for attention!

Anna Trubetskaya
Technical University of Denmark
Chemical Engineering Department
Combustion and Harmful Emission Control Group
Søltofts Plads, Bygning 229
2800 Lyngby, Denmark
***email:* atru@kt.dtu.dk**

DTU Chemical Engineering
Department of Chemical and Biochemical Engineering

427217 Small-Scale Biomass Gasification - Research on Entrained Flow and Fluidized Bed Gasification Technology

Abstract:

The increasing demand for safe, sustainable and affordable renewable energy today promotes development of small-scale biomass gasification systems for power and fuel production in Europe. Different technologies, process designs and conversion concepts are developed focusing on entrained flow and fluidized bed gasification technology. Comparing the technologies concerning efficiencies, economics, technological readiness and alternative process chains, research and development needs are carved out and translated to current research projects at the TUM.

At the Institute for Energy Systems, we conduct gasification experiments on entrained flow and fluidized bed biomass gasification. The test rigs for bubbling fluidized bed gasification and entrained flow gasification can handle various biogenic feedstock with a thermal input up to 100 kW_{th} and temperatures up to 1600°C. Furthermore, elaborate fuel and gas analysis measures ensure the evaluation of the gasification processes at a scientific level. Additional process simulation studies and CFD modelling enable process evaluation and include pre-treatment and product gas upgrading options.

Specific areas of research are hydrothermal carbonization (HTC) of biomass as pre-treatment technology to enable entrained flow gasification of diverse biogenic feedstock. Hot gas cleaning technology for tar and hydrocarbon reforming as well as adsorptive trace elements removal are tested in long-term experiments conducting parameter studies downstream of the fluidized bed gasification. Product gas methanation as example for synthesis gas utilization is investigated concerning catalyst deactivation and poisoning as well as concerning the complex heat removal from the exothermic reaction.

433590 Catalytic Effect of Na, K, Ca and Mg on CO2 Enhanced Gasification of Biomass Char

Abstract:

Biomass can be converted to biofuels using various pathways which are broadly classified as biochemical or thermochemical. Gasification is a common thermochemical process that converts biomass to a gaseous product, syngas, which can be used for power generation, chemical production and as a fuel. The use of air, steam and oxygen as oxidizing agents in this process has been widely studied and reported. Catalytic gasification of biomass can effectively improve the reaction rate and conversion efficiency compared to conventional gasification. However, very little is known about the catalytic gasification of biomass char with CO_2.

In the current study, unloaded biomass char was gasified using CO_2 at five different temperatures between 800°C – 950°C in a fixed bed reactor. Using the evolution of CO in the gasification process as basis for conversion, the conversion-time data was fitted to three different solid-gas reaction models. The Random Pore Model (RPM) gave the best fit and the kinetic parameters for the same were calculated. In the second part of the study, acetates of Sodium (Na), Magnesium (Mg), Potassium (K) and Calcium (Ca) are used to load the char with metal catalysts and subjected to the same gasification procedure. The effect of catalysts is studied using the variation in kinetic parameters obtained using the RPM.

Instruments such as a CHNS analyzer, inductively coupled plasma (ICP), Scanning Electron Microscope (SEM) are used to analyze the composition and properties of char. Additionally, a micro GC is used to analyze the syngas at the output.

425799 Co-Gasification of Sugarcane Bagasse with Cane Leaves/Tops: Role of Potassium Migration on Char Reactivity

Abstract:

Co-gasification of biomass with petroleum coke is known to enhance the reactivity of less reactive petroleum coke by the alkali and alkaline earth metals present in a more reactive biomass char (synergy). Since biomass species differ in their ash content and composition, there is a possibility of enhancement in the reactivity of less reactive biomass char by a more reactive biomass char during co-gasification of two different types of biomass. Also, an added incentive for co-gasification of two or more biomass is the potential improvement in the economy of scale and fuel availability. In this study, the focus is on co-gasifying two kinds of sugarcane residue, bagasse and sugarcane leaves/tops (SCT) chars. The Brazilian bagasse used in this study has a much higher content of potassium, and thus much higher overall char reactivity, than the Brazilian SCT char. Thus, co-gasification was expected to improve the reactivity of SCT char by the synergistic effect, which is the main focus of this study.

Contrary to expectations, the experimental results during co-gasification of bagasse and SCT chars showed that for conversions below 75%, the observed reactivity of the mixture is lower than the predicted reactivity based on no interaction between bagasse and SCT chars (inhibition). When the char conversion exceeds 75%, the ash from bagasse char starts to enhance the reactivity of SCT char in a mixture (synergy). The initial inhibition effect is attributed to migration of potassium from K-rich bagasse char to SCT char, followed by the reaction of some of the migrated potassium with silicates in SCT char to form inactive potassium aluminosilicates. Potassium migration was confirmed by SEM-EDX technique. Addition of SCT ash to bagasse char lowered the reactivity of bagasse char, while the addition of bagasse ash to SCT char enhanced the reactivity of SCT char. Based on these results, different flow schemes for co-processing of bagasse and SCT chars are suggested to maximize the char reactivity.

435355 Performance of 3 ton/day BFB Gasification System using Pine Feedstock

Abstract:

Partial oxidation of carbonaceous fuels at high temperatures (gasification) is a process to produce alternative energy. On promising technique to produce alternative energy is Biomass gasification. The alternative energy facility of University of Louisiana at Lafayette in collaboration with Cleco Power LLC., has implemented a 3 ton/day bubbling fluidized bed gasification system to operate and collect data using both air and oxygen as oxidizing medium, and produce syngas to generate electricity.

The effect of gasification temperature, equivalence ratio (ER), and steam injection on syngas compositions, gasifier efficiency and tar composition were studied. We have demonstrated that pilot scale gasification of raw pine achieving around 8% of Hydrogen and 10% of Carbon monoxide in syngas sufficient for running 25 kW AC generator. Raw pine wood was gasified with natural air with/without steam in a bubbling Fluidized Bed-Reactor (FBR) with crystalline silica quartz as catalyst at bed temperature of 1550°F. The equivalence ratio (ER) was maintained at 0.3 and material was feed at 120 lb/hr. The syngas had a heating value (HHV) of 107.79 Btu/scf and gas yield 78scf/lb. Higher temperature increased Hydrogen & Carbon Monoxide production and carbon conversion. Tar must be removed to increase the life of the bag filter and heat exchanger tubes. Another challenge is that water accumulation in the system due to high moisture content in the raw pine chips. At present, gasification of torrefied pine is being evaluated to increase the heating value of the syngas. The efficiency of the combine torrefaction –gasification process of would depend on the yield of the torrefaction operation.

Investigation of Most Effective Parameters on Biomass Gasification in Circulating Fluidized bed Gasifiers

Guilnaz Mirmoshtaghi
Jan Skvaril
Hailong Li
Eva Thorin
Erik Dahlquist

2015 AIChE Annual Meeting, Salt Lake City, UT

Introduction

- Biomass is a renewable resource of energy for fuel, heat and power production
- One of the technologies to convert the solid fuel energy to usable energy is gasification.
- In gasification, different steps and reactions are interrelated
- A system with high level of complexity
- Still unknown relations between the gas composition and operating parameters.
- In some cases, **specific component** or specific operating condition
- In some cases **air gasification** of a **specific biomass**
- In some cases **different biomass species** in one specific gasifier

2

Objective

1. **Defining the relation between input and output data and analyze the effect of different operating** parameters
   a. Is there any specific criteria for any gasifier?
   b. Is there any correlation between input and output data?
   c. What is the impact of different operating conditions (input data) on the **carbon conversion**, **tar yield** and **heating value of the product gas**?
2. **What are the important parameters** for production of different components in the product gas?

3

Materials and Methods

- Principle component analysis (PCA)
  - Explorative data analysis
  - Exctract information, noise removal, dimensionality reduction

X	=	Model	+	Error
Data	=	Structure	+	Noise
X	=	T P^T	+	E

- Each variable defined on variable space

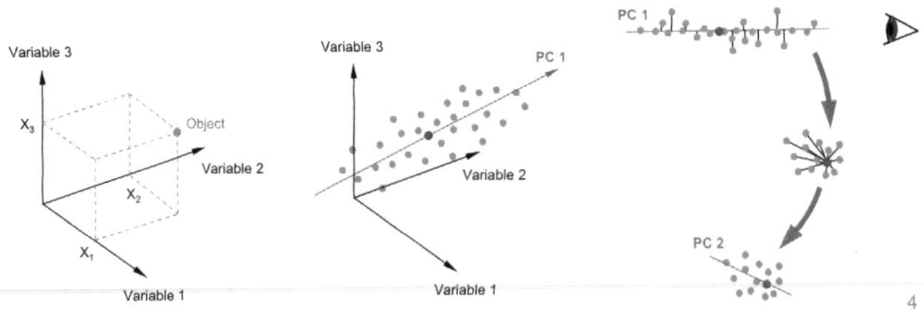

4

Materials and Methods

- Scores
  - Show how well the data are distributed, sample patterns, groupings, similarities and differences

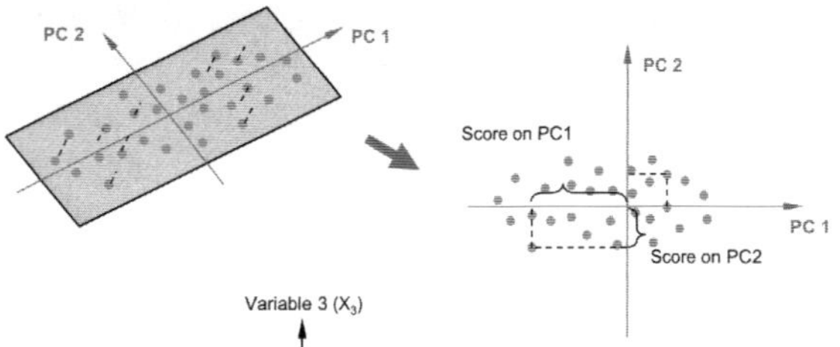

- Loadings
  - Every variable has a loading on each PC, which reflects how much the individual variable contributes to that PC
  - A loading is the cosine of the angle between the variable and the current PC, the smaller the angle the larger the loading
  - Useful to understand the correlations between variables

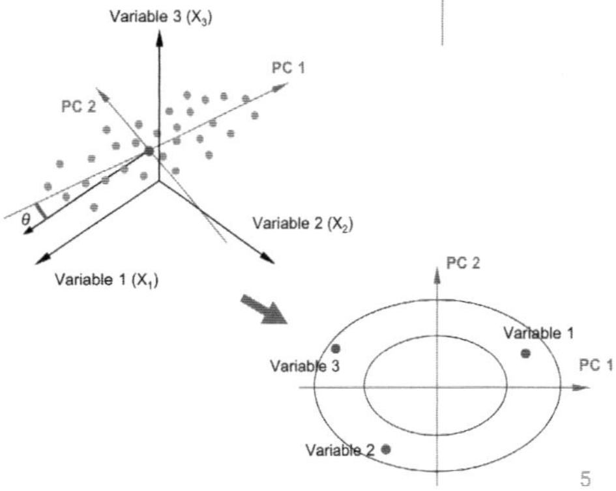

5

Materials and Methods

- PLS
  - Partial Least Squares Regression or Projection to Latent Structures
  - PLS aims to maximize the covariance between X and Y
  - Scores and X and Y loadings
- Regression coefficients
  - show how each variable is weighted when predicting a particular Y response.
- Uncertainty testing of regression coefficients
  - Requires cross-validation
  - jack-knifing
  - Giving information about importance of variables

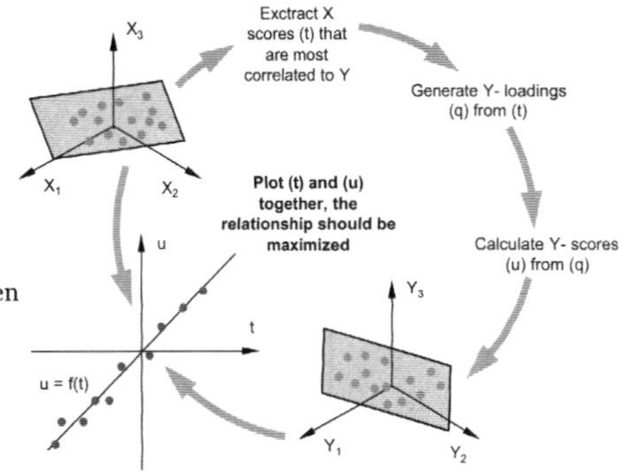

6

Materials and Methods

Gasifiers	Abbreviation	Number of data points	Reference
BIVKIN gasifier-ECN company	ECN	11	(Van der Drift et al., 2001)
TPS company	TPS	13	(Waldheim et al., 2002)
Li -University of British Columbia	Li	14	(Li, X.,2001)
Wu Jianzhi - Guangzhou institute of energy conversion	Wu	6	(Wu, J. et al., 1992)
Garcia-CIEMAT institute	Garcia	4	(Garcia-Ibanez et al., 2002)
VTT institute	VTT	10	(Esa Kurkela et al, 1991-1992)
Qi Miao-University of western Ontario	Miao	14	(Miao et al., 2013)
Christodoulou-Chemical process and energy resources institute, Hellas	Chris	12	(Christodoulou et al., 2014)
X-Meng- Delft university of Technology	Meng	8	(Meng et al., 2011)

7

Materials and Methods

Variables	Unit	Observations	Min	Max	Average	STD*
			Independent variables			
C	Wt%	92	39.78	54.4	47.5	4.32
H	Wt%	92	4.97	7.86	6.01	0.74
O	Wt%	92	30.7	48.9	40.09	3.43
Ash	Wt%	92	0.14	47.3	7.5	8.6
Moisture	Wt%	92	3.5	22	9.8	3.74
ER	-	92	0.17	0.73	0.32	0.1
Temperature	°C	92	681	1042	804	67.28
S/B**	-	92	0	9	0.85	2.34
B/V***	kg/m^3.hr	92	40.14	950	351.78	201.95
HHV biomass	MJ/kg	92	14.14	25.8	19.4	3.23
			Dependent variables ****			
H_2	Vol%	92	**7.28**	**31.1**	**19.07**	**5.73**
CH_4	Vol%	92	**3.4**	**20.3**	**8.91**	**2.67**
CO	Vol%	92	**19.06**	**46.42**	**30.71**	**6.62**
CO_2	Vol%	92	**24.58**	**65.13**	**40.15**	**8.1**
C_2H_4	Vol%	92	**0**	**4.25**	**1.33**	**1.3**
LHV product gas	MJ/m^3	92	1.66	7.4	4.32	1.42
Carbon conversion	-	92	57.5	100	87.4	8.73
Product gas flow	m3/hr	92	13.82	2281.65	584.9	704.6
Dry gas yield	m3/kg	92	1.31	6.26	2.3	0.73
Tar yield	gr/nm3	92	0.04	15.13	4.3	2.78

* Standard deviation

**Steam mass flow to biomass flow ratio

*** Biomass flow to volume of the reactor

**** The blue numbers are the results after normalization

- Data were mean centered and normalized

8

Results

- **Is there any specific criteria for any gasifier?**

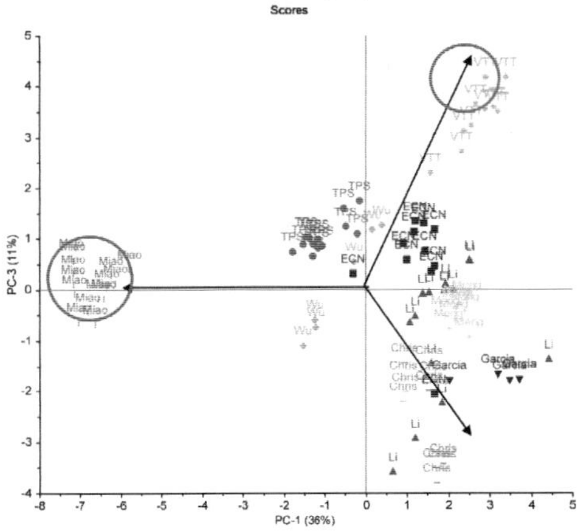

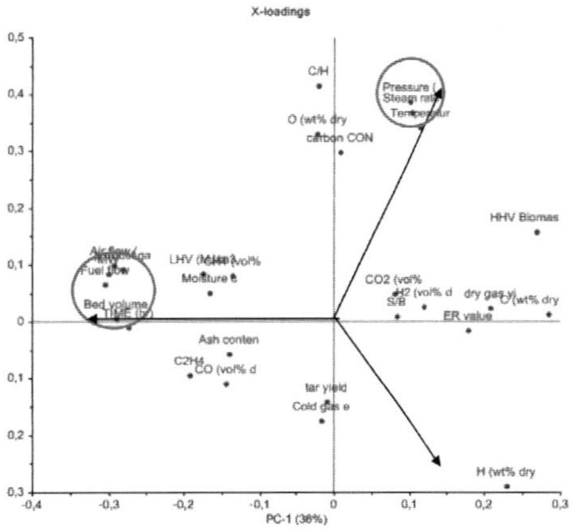

- VTT is characterized by high pressure, steam rate and temperature
- Miao is characterized by large Volume and residence time

9

Results

- **Is there any specific criteria for any gasifier?**

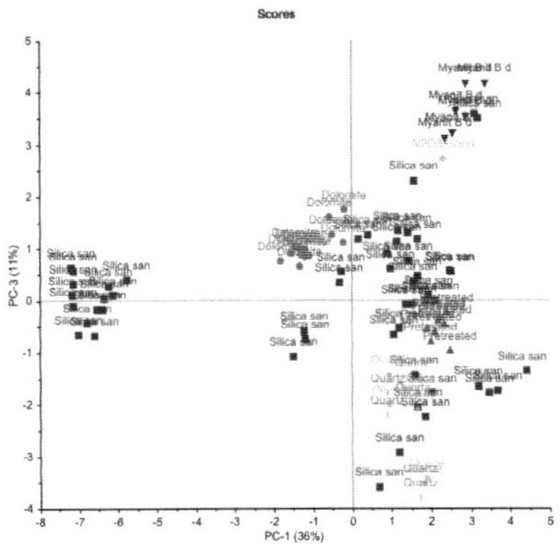

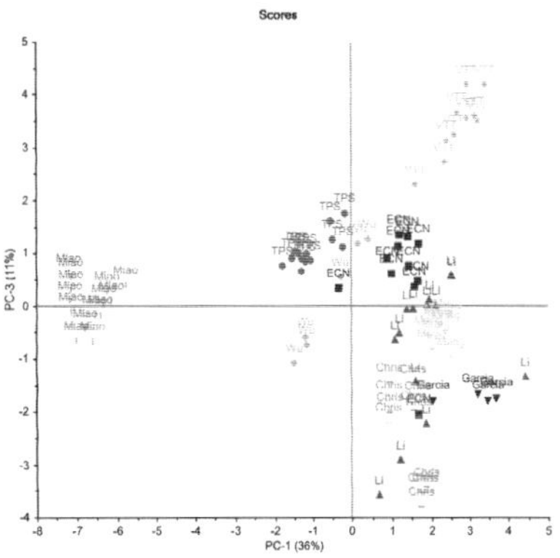

- Miao, ECN, Garcia, Li and Wu and partly VTT are specified by Silica sand as their bed material
- TPS has Dolomite as the bed material
- Chris and Meng are operating on Quartz and pretreated Olivine, respectively
- VTT is partly operated on a bed of Myanit B dolomite and Al_2O_3

10

Results

- **Is there any correlation between input and output data?**

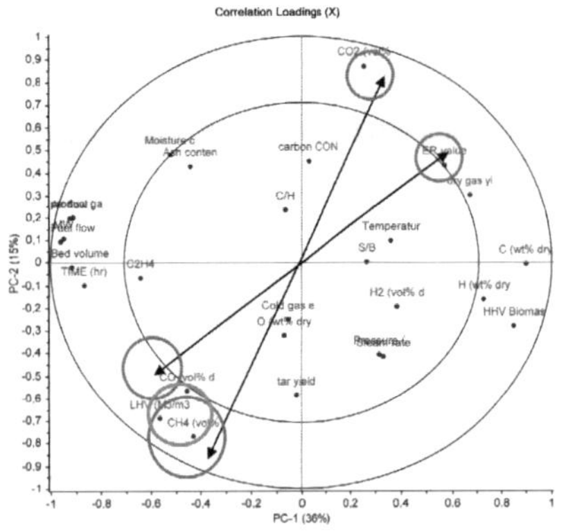

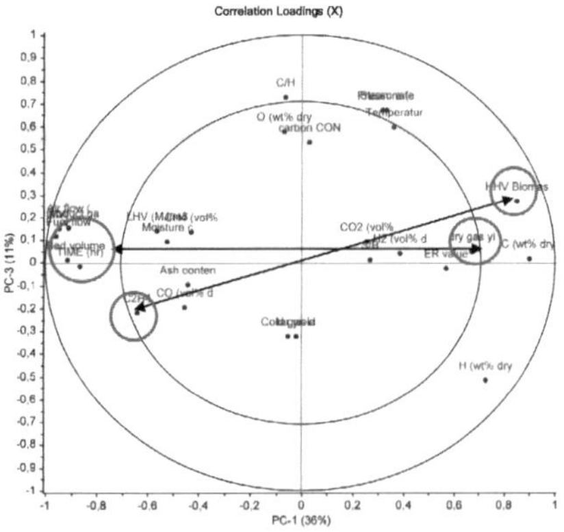

- Negative correlation between ER value and CO concentration
- Negative correlation between CH_4 and CO_2 concentration
- Positive correlation between CO, CH_4 concentration and LHV of the product gas
- Negative correlation between dry gas yiled and the gasifier voulme
- Negative correlation between HHV biomass and C_2H_4 concentration

11

Results

- **Is there any correlation between input and output data?**

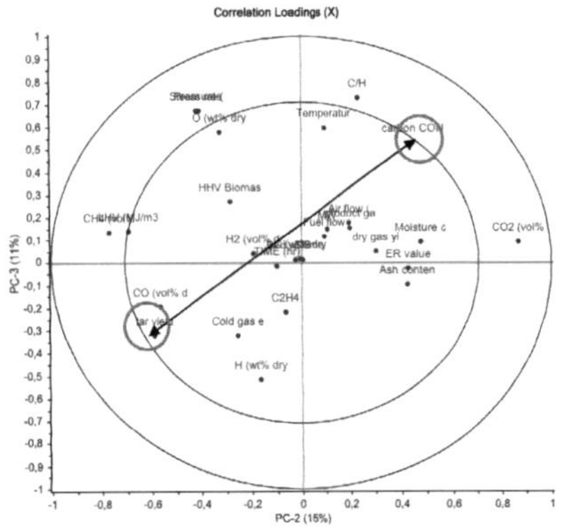

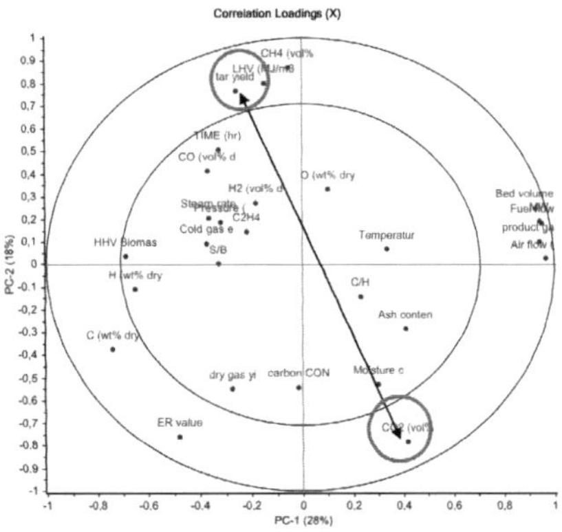

- Negative correlation between tar yield and carbon conversion
- Negative correlation between tar yield and CO_2 concentration

12

Results

- **Is there any correlation between input and output data?**

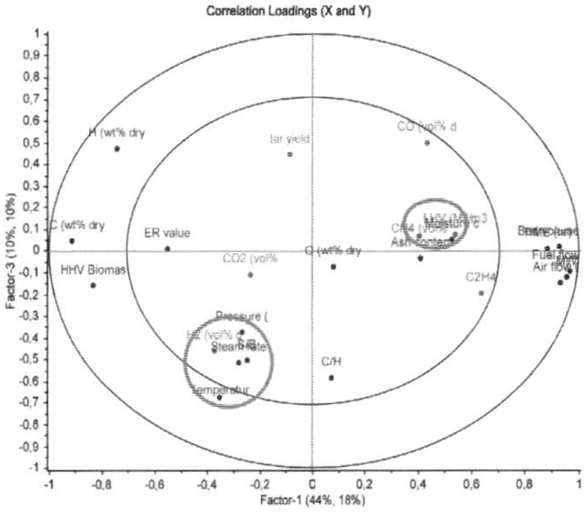

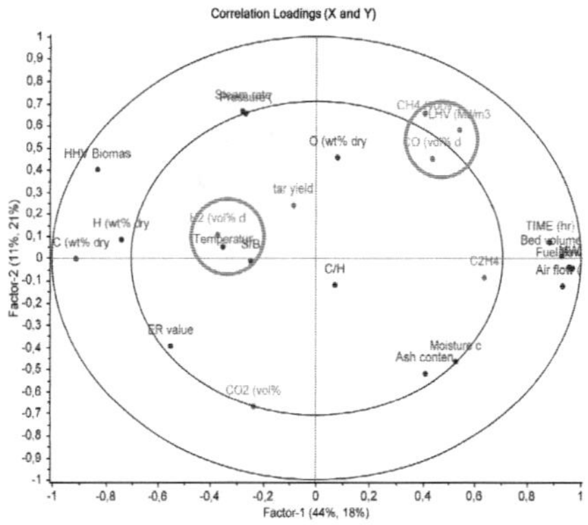

- Positive correlation between H_2 concentration and steam rate, S/B, pressure and temperature
- Positive correlation between CH_4 concentration, LHV of the product gas

13

Results

- **What are the important parameters for production of different components in the product gas?**

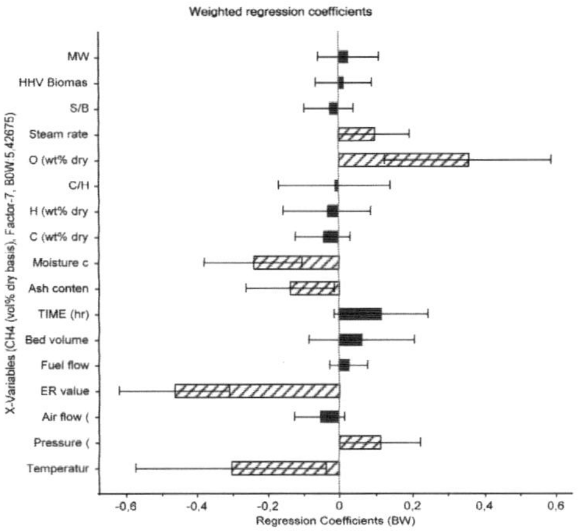

Parameters affecting CH_4 concentration

- Strong positive effect by steam rate, pressure and oxygen content in the biomass
- Strong negative effect by Moisture content, ash content, temperature and ER value

In order to produce **more CH_4**:

- A biomass with low moisture content, ash content and high oxygen content
- Operation of the reactor in low ER values, low temperature, high pressure and high steam rate

14

Results

- **What are the important parameters for production of different components in the product gas?**

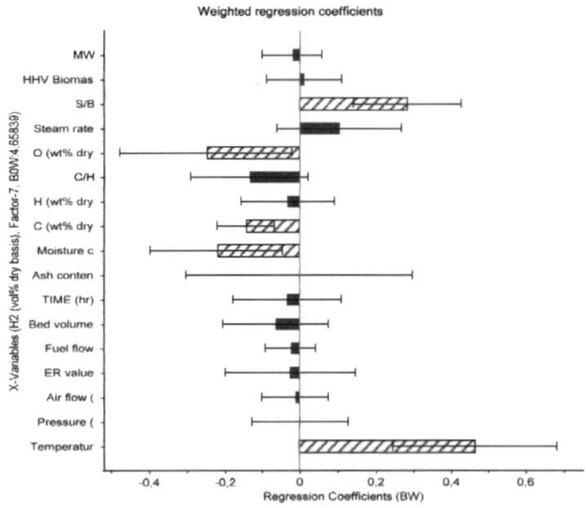

Parameters affecting H_2 concentration

- Strong positive effect by S/B and temperature
- Strong negative effect by Moisture content and Oxygen and carbon content in the biomass

In order to produce **more H_2**:

- A biomass with low mositure content and low oxygen and carbon content
- Operation of the reactor in high temperature and S/B ratio

15

Results

- **What are the important parameters for production of different components in the product gas?**

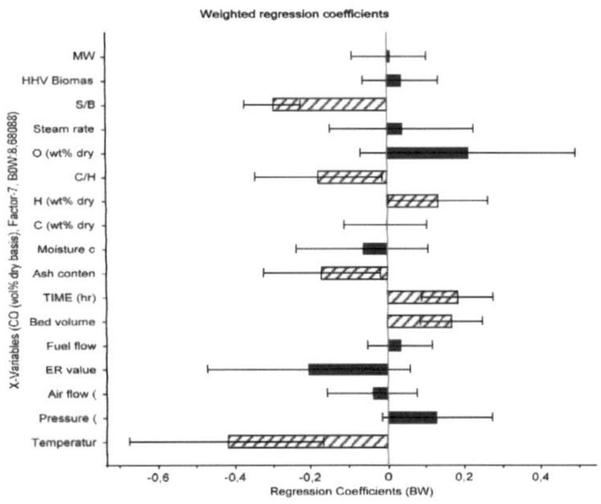

Parameters affecting CO concentration

- Strong positive effect by hydrogen content in the biomass, residence time and reactor volume
- Strong negative effect by temperature, S/B ratio, C/H ratio and ash content of the biomass

In order to produce **more CO**:

- A biomass with high hydrogen content and low C/H ratio and low ash content
- A large gasifier with long residence time
- Operation of the reactor in low temperature and low S/B ratio

16

Results

- **What are the important parameters for production of different components in the product gas?**

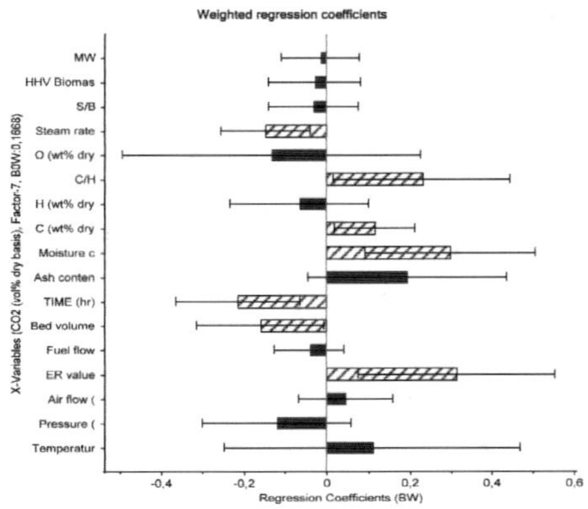

Parameters affecting CO_2 concentration

- Strong positive effect by ER value, Mositure content, carbon content and C/H ratio in biomass
- Strong negative effect by steam rate, bed volume and residence time

↓

In order to produce **less CO_2**:

- A biomass with low mositure content, low carbon content and low C/H ratio
- Operation of the large reactor with high residence time by low ER value and high steam rate

17

Results

- **What is the impact of different operating conditions (input data) on the heating value of the product gas?**

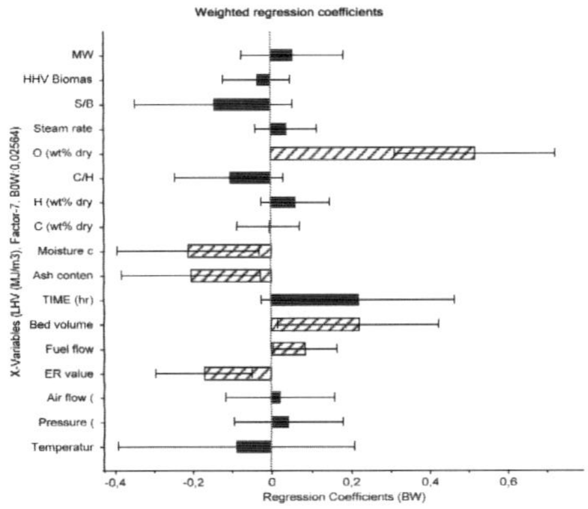

Parameters affecting LHV of the gas

- Strong positive effect by oxygen content of biomass, bed volume and fuel flow rate
- Strong negative effect by ER value, mositure content and ash content of biomass

↓

In order to have gas with **higher LHV**:

- A biomass with high Oxygen content, low moisture and ash content
- Operation of the large reactor with long residence time by high fuel flow and low ER value

18

Results

- **What is the impact of different operating conditions (input data) on the tar yield?**

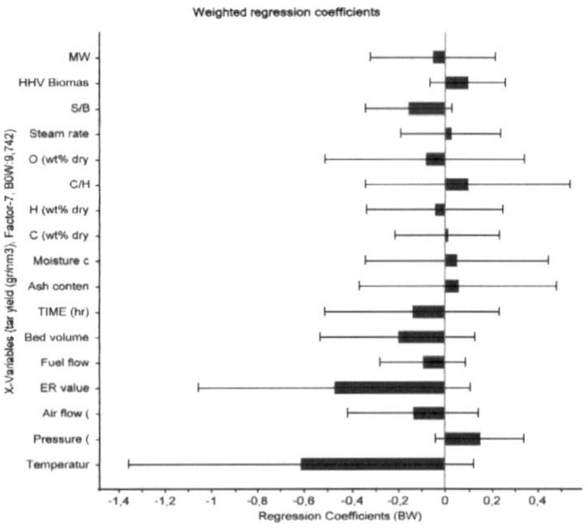

Parameters affecting Tar yield

- Negative effect* by ER value and temperature

↓

In order to have gas with **lower tar yield**:

- Operation of the reactor by high ER value and high temperature

*** According to the model, none of the input parameters are siginificantly affecting tar yield, however these two parameters show the largest coefficient**

19

Results

- **What is the impact of different operating conditions (input data) on carbon conversion?**

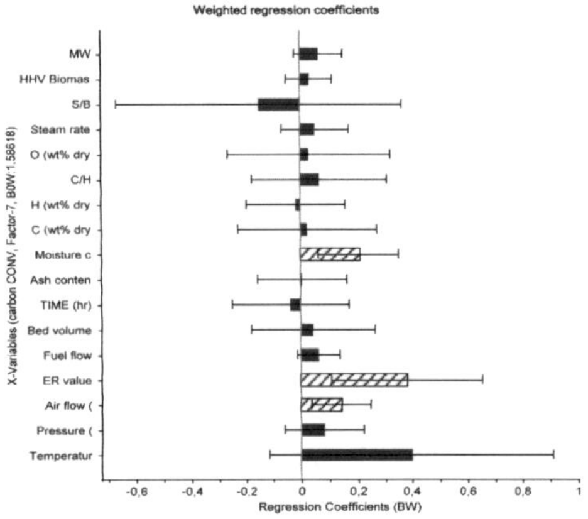

Parameters affecting Carbon conversion

- Strong positive effect mositure content in biomass, ER value and air flow rate

↓

In order to have gas with **higher carbon conversion**:

- A biomass with high moisture content
- Operation of the reactor by high ER value and high air flow rate

20

Conclusions

- CH_4 is produced **more** if :
  - A biomass with **low mositure content**, **low ash content** and high oxygen content is used.
  - The gasifier is operated in low ER values, low temperature, high pressure and **high steam rate**.

- H_2 is produced **more** if:
  - A biomass with **low mositure content** and low oxygen and low carbon content is used.
  - The gasifier is operated in high temperature and **high S/B ratio**.

- CO is produced **more** if:
  - A biomass with high hydrogen content, low C/H ratio and **low ash content** is used.
  - The **large gasifier** with **long residence time** is used.
  - The gasifier is operated in low temperature and low S/B ratio

- CO_2 is produced **less** if:
  - A biomass with **low mositure content**, **low carbon content** and **low C/H ratio** is used.
  - The **large gasifier** with **long residence time** is operated in low ER value and **high steam rate.**

21

Conclusions

- In order to have gas with **larger** LHV:
  - A biomass with **low mositure content**, high Oxygen content and **low ash content** is used
  - The **large gasifier** with **long residence time** should be operated by high fuel flow and low ER value

- In order to have **lower** Tar yield:
  - The gasifier should be operated at high temperature and high ER value

- In order to have **higher** Carbon conversion:
  - A biomass with high moisture content is used
  - The gasifier should be operated in high ER value and high air flow rate

22

Acknowledgment

The authors express their gratitude to :

- **Swedish energy agency** and
- **Swedish gasification center**

for major financing of this project.

23

Thank you for your attention

Any Questions?

Guilnaz Mirmoshtaghi
Mälardalen University
Västerås-Sweden

Tel: +46-21-101490
Email: guilnaz.mirmoshtaghi@mdh.se

437230 Gasification of Maize Cob Catalyzed By Modified Iron-Rich Red Mud Extract

Abstract:

Gasification of biomass gives the synthesis gas which may either be further processed to give Fischer-Tropsch diesel fuel or burn for electricity. At present, one of the most used kinds of catalysts for the biomass gasification are alkali metals, such as iron, potassium and sodium, which facilitates the formation of gaseous production especially hydrogen, and also promotes the release of gases tended to happen at a lower temperature. However, whether the catalyst was added in an impregnation way or dry mix way, the recovery is difficult, which increases the cost of gasification. The Bayer red mud is a highly available iron-containing material with low or no cost, but there are few reports about using the red mud extract as the catalysts of gasification. In this paper, different ratios of modified red mud extract were impregnated into the maize cob, and the oxygen was used as gasifying agent in a fixed bed gasifier. The catalytic effect caused by the extract came from same amount of red mud with maize cob was same as the nitrate iron in 5 wt%.The modifications of the red mud with acid improving its surface area activity and changing its compositions might be the possible reason. Using red mud extract as catalyst will be a great way both to improve gasification of biomass and to handle hazardous red mud.

422351 Integration and Chemical and Fuel Intermediate Production from Post-Consumer Municipal Solid Waste

Abstract:

Integration and Chemical and Fuel Intermediate Production from Post-Consumer Municipal Solid Waste

Ling Liang[1], Chenlin Li[1], Qian He[1], Marcello Somma[2], Nicola D'Alessio[2], Ning Sun[1], Todd R. Pray[1]

[1]Advanced Biofuels (and Bioproducts) Process Demonstration Unit, Lawrence Berkeley National Laboratory, Emeryville, CA, United States

[2] FATER S.p.A, Pescara, Italy

Abstract

As populations grow and consumer packaged goods (CPG) consumption increases worldwide, the re-utilization and valorization of municipal solid waste (MSW) will become increasingly critical. Product and market development using this abundant feed-stream will require process optimization, integration, validation and scale-up demonstration and prototyping to mitigate environmental impact and carbon emissions. Incineration or landfill disposal is not a sustainable practice in the long term, and simple recycling may not always be feasible due to waste stream composition and logistics.

To address this global environmental sustainability and economic issue FATER, a leading European CPG supplier and joint venture of Procter & Gamble and the Angelini Italian industrial group, is collaborating with the Advanced Biofuels *[and Bioproducts]*Process Demonstration Unit (ABPDU) at the Lawrence Berkeley National Laboratory. The ABPDU has been funded and managed by the US Department of Energy's Bioenergy Technologies Office (BETO) within the Office and Energy Efficiency and Renewable Energy (EERE).

This presentation will highlight the process and product development FATER and the ABPDU have teamed on. We will provide an overview and strategic insights into the novel technologies, markets and molecules accessible via efficient and value-added waste stream utilization. This unique public-private partnership demonstrates how biomass and MSW conversion and upgrading via bioprocessing can influence key pathways to address global environmental sustainability barriers.

427791 Exploring the Impact of Cellulose Accessibility on Lignocellulosic Hydrolysis Using a Modified Simons' Staining Assay and a 1-Methylimidazole Pretreatment Process

Abstract:

To understand and enhance cellulose biodegradation, it is crucial to determine how a pretreatment method affects cellulose accessibility. Simons' staining is a simple technique to quantify cellulose accessibility using direct dyes. However, the uncertainty around the availability of traditional dyes, Direct Orange 15 and Direct Blue 1, has created the need for alternative probes. In this work, we demonstrate the use of Direct Yellow 11 as an alternative probe in a modified Simons' staining method. With this modified Simons' staining method, we are able to characterize the accessibility of a 1-Methylimidazole (MI) pretreatment process. MI is a potential biomass pretreatment chemical that removes lignin and dramatically increases hydrolysis rates at ambient conditions and short times. Here we elucidate how MI pretreatment affects both the crystallinity and accessibility of cellulose. This work provides measurable criteria for other organic solvent pretreatments, and it will guide future efforts towards environmental, economical, and effective pretreatment methods.

(1) Simons, F. L. (1950). A stain for use in the microscopy of beaten fibers. Tappi, 33(7), 312-314.

(2) Kang, Y., Realff, M. J., Sohn, M., Lee, J. H., & Bommarius, A. S. (2015). Biotechnology Progress, 31, 25-34.

(3) Chandra, R., Ewanick, S., Hsieh, C., & Saddler, J. N. (2008). Biotechnology progress, 24(5), 1178-1185.

434700 Effects of Multiple Inhibitions in Corn Cob Hydrolysate on the Lipid Production of Rhodotorula Glutinis

Abstract:

Effects of multiple inhibitions in corn cob hydrolysate on the lipid production of Rhodotorula glutinis

Hongjuan Liu*, Zhangnan Lin, Jianan Zhang*

Institute of Nuclear and New Energy Technology, Tsinghua University, Beijing 100084, China. Email: liuhongjuan@tsinghua.edu.cn or zhangja@tsinghua.edu.cn

Abstract: As an alternative feedstock for biodiesel production, microbial lipid production with cheap renewable resources such as lignocellulosic biomass has attracted people's attention. However, lignocellulose degradation byproducts that generated in the pretreatment process have various negative impacts on cell growth and lipid production. Ordinarily, the effects of the multiple inhibitors are synergy interaction. In this work, multiple inhibitions of acetic acid, furfural and 5-hydroxymethylfurfural (HMF) on the cell growth and lipid production of oleaginous yeast Rhodotorula glutinis with corn cob hydrolysate was studied. Among the three inhibitors, furfural is toxic most for the cells, followed by HMF and acetic acid. Rhodotorula glutinis has the strong tolerance to acetic acid. The cell growth and lipid production could even be stimulated when acetic acid concentrations were lower than 10 g/L. Synergy inhibition of furfural and acetic acid on Rhodotorula glutinis was significant, but it was not obvious between acetic acid and HMF or furfural and HMF. When acetic acid concentration higher than 4g/L combined with furfural concentration above 0.2 g/L, the biomass and lipid production declined dramatically. The results were useful for the deep understanding of Rhodotorula glutinis metabolism and helpful for the optimization of lipid production with liginocellulosic biomass.

Key words: liginocellulosic biomass, microbial lipid, inhibition, fermentation, Response surface analysis

434285 Development of a Blended Feedstock Strategy to Address Challenges of Feedstock Cost for Biochemical Conversion of Lignocellulosic Biomass

Abstract:

A blended feedstock strategy is currently under investigation at Idaho National Laboratory to address challenges associated with the high cost of feedstock access in order to enable biomass supply systems. Blending provides a mechanism to reduce risk to the supply chain and reduce feedstock cost by coupling location-specific resource utilization with biofuel production. Using corn stover as a baseline, this study examines a variety of biomass resources, including switchgrass, *Miscanthus*, wheat straw, non-recyclable paper, lawn clippings, energy cane, and sorghum to formulate feedstock blends for biochemical conversion. The objectives of this work are: 1) to characterize and understand the effects of blending and logistics on the efficacy of pretreatment and conversion of single feedstocks and blends; 2) to examine the potential for blends to reduce feedstock production costs using least cost formulation models and feedstock supply chain technoeconomic analysis. Feedstocks and blends were screened using an automated dilute-acid pretreatment method combined with enzymatic hydrolysis; sugar yields were used to develop conversion performance relationships among single feedstocks and blends. Physical and chemical characteristics before and after pretreatment were also used to evaluate the conversion performance of biomass blends. In order to enable biomass supply chains and achieve biochemical conversion targets, feedstock blending has shown promise as a strategy to supply biomass in quantities sufficient for meeting cost, quality, sustainability, and performance targets.

434921 Biogas Production Trend at Various Temperature Conditions and Methane Enrichment Using Solvent Absorption Method

Abstract:

Depletion of the fossil fuels and their prices are going up therefore it is the need of the hour to find out alternative renewable energy resources. At the same time the garbage production is increasing and creating serious environmental and health risks. This waste requires safer treatment to maintain human health and natural environment. Nowadays, main focus is on transformation of waste into value added products as well as energy production. This work investigates the biogas production considering the effect of pH and temperature. Methane enrichment process by removal of CO_2 & H_2S was also the part of this study. Fruit and vegetable waste (FVW) of banana, apple, orange, potato, poes peels and cabbage were used as raw material for biogas production. Lab scale batch anaerobic digestion (AD) is conducted at mesophilic temperature (MT) 35°C and thermophilic temperature (TT) 55°C. The raw biogas is pebbled through water, NaOH, $Ca(OH)_2$ and triethanolamine (TEA) for purification.

The experimental results show that mixed fruit waste (MFW) provides 10% more biogas yield than mixed fruit vegetable waste (MFVW). The maximum biogas yield was 0.030 (g/g volatile solids) at TT. The optimum pH range is found between 8 -8.5. The use of chemical solutions like NaOH, $Ca(OH)_2$ and TEA instead of water increases methane enrichment upto 5%, 9% and 7% respectively. The impurities CO_2 and H_2S decreased upto 28% and 150 ppm respectively by using $Ca(OH)_2$. Process is effective to enhance the gas production with change of process conditions and enrichment of biogas by removing carbon dioxides and hydrogen sulfide.

417241 Using Microbial Consortia to Increase the Productivity of Algal Open Ponds

Abstract:

In March 2015, the United States pledged to cut its carbon emissions 26-28% by 2025 [1]. To achieve this goal, the development of cheap biofuels is necessary. First generation biofuels are obtained from food crops. The realization that these biofuels are only slightly better than traditional fossil fuels regarding environmental impact combined with their competition for food resources prompted research on second-generation biofuels which are obtained from waste biomass [2]. The most promising technologies for second-generation biofuels production are based on microbial conversion of biomass into lipids. Preferred characteristics of the microorganisms are high specific growth rate, high lipids to biomass yield, high cell density, ability to use complex substrates, affinity to substrate, and low nutrient requirements [3]. These characteristics are found in some strains of microalgae.

Microalgae are attractive because some strains accumulate up to 50% dry weight in lipids [4]. Also, algae do not compete for food resources, they can be grown on wastewater and/or seawater, and they are up to one order of magnitude more efficient than higher-order terrestrial plants in capturing sunlight [4], [5]. Despite their huge promise in reducing greenhouse gas emissions, algal biofuels are not yet economically feasible as closed photobioreactors incur high capital and operating costs, and cheap open pond systems have low biomass and lipids productivities. Open ponds have low productivities because they are CO_2 limited and they are susceptible to invasion by undesirable microorganisms. In this paper we introduce the idea of using microbial consortia to counteract these limitations.

Traditionally, CO_2 limitations have been addressed by sparging flue gas [6], [7], and no real solution has been provided to invasion. Microbial consortia can address both limitations. First, microbial consortia provide resilience by utilizing resources more efficiently compared to an algal monoculture, making pond invasion by undesirable species more difficult [8]. Second, symbiotic relationships can be established between microalgae and other microbes in which the microbes convert carbon sources into CO_2 and algae absorbs the CO_2 into biomass. Therefore, algae ponds can become feasible in locations far away from flue gas sources.

The lack of predictive and reliable models of dynamic bioprocesses has halted the exploration of novel strategies to improve productivities, such as microbial consortia. Dynamic flux balance analysis (DFBA) provides a way of exploiting equipment process models and genome-scale metabolic networks to make predictions of novel system behavior [9], [10], [3]. Recently published simulators enable reliable and efficient simulation of DFBA systems [11], [12]. In this paper we use the High-Rate Algal Pond model [13], [14] and DFBA to model an algae/yeast coculture and explore its benefits. Results show that microbial consortia can make algae ponds economically feasible in locations far away from flue gas sources, but with available agricultural waste.

Keywords: Algae pond, microbial consortia, raceway pond, dynamic flux balance analysis, synthetic ecology.

References

[1] R. Harrabin, "US makes climate pledge to UN," BBC News, 31 March 2015. [Online]. Available: http://www.bbc.com/news/science-environment-32136006.

[2] R. Luque, "Algal biofuels: the eternal promise?," *Energy & Environmental Science,* vol. 3, pp. 254-257, 2010.

[3] K. Höffner and P. I. Barton, "Design of Microbial Consortia for Industrial Biotechnology," *Computer-Aided Chemical Engineering,* vol. 34, pp. 65-74, 2014.

[4] M. S. Wigmosta, A. M. Coleman, R. J. Skaggs, M. H. Huesemann and L. J. Lane, "National microalgae biofuel production potential and resource demand," *Water Resources Research,* vol. 47, p. W00H04, 2011.

[5] Y. Chisti, "Constraints to commercialization of algal fuels," *Journal of Biotechnology,* vol. 167, pp. 201-214, 2013.

[6] J. Doucha, F. Straka and K. Lívanský, "Utilization of flue gas for cultivation of microalgae Chlorella sp.) in an outdoor open thin-layer photobioreactor," *Journal of Applied Phycology,* vol. 17, no. 5, pp. 403-412, 2005.

[7] L. M. Brown, "Uptake of Carbon Dioxide from Flue Gas by Microalgae," *Energy Convers. Mgmt.,* vol. 37, no. 6-8, pp. 1363-1367, 1996.

[8] E. Kazamia, D. C. Aldridge and A. G. Smith, "Synthetic ecology - A way forward for sustainable algal biofuel production?," *Journal of Biotechnology,* vol. 162, pp. 163-169, 2012.

[9] J. D. Orth, I. Thiele and B. Ø. Palsson, "What is flux balance analysis?," *Nature Biotechnology,* vol. 28, pp. 245-248, 2010.

[10] R. Mahadevan, J. Edwards and F. I. Doyle, "Dynamic flux balance analysis of diauxic growth in Escherichia coli.," *Biophysical Journal,* vol. 83, no. 3, pp. 1331-40, 2002.

[11] K. Höffner, S. M. Harwood and P. I. Barton, "A reliable simulator for dynamic flux balance analysis," *Biotechnology and Bioengineering,* vol. 110, no. 3, pp. 792-802, 2013.

[12] J. A. Gomez, K. Höffner and P. I. Barton, "DFBAlab: A fast and reliable MATLAB code for Dynamic Flux Balance Analysis," *BMC Bioinformatics,* vol. 15, p. 409, 2014.

[13] H. Buhr and S. Miller, "A Dynamic Model of the High-Rate Algal-Bacterial Wastewater Treatment Pond," *Water Res.,* vol. 17, pp. 29-37, 1983.

[14] A. Yang, "Modeling and Evaluation of CO2 Supply and Utilization in Algal Ponds," *Industrial & Engineering Chemistry Research,* vol. 50, pp. 11181-11192, 2011.

424078 Algal Culture with a Nitrogen Fixing Bacterium for Sustainable Biomass Production

Abstract:

Macronutrients such as nitrogen and phosphorus are directly tied to biomass yields. There is currently a significant and growing interest in next generation biofuel targets such as algae, which promise higher biomass yields in combination with increased biofuel production. Yet, these increased biomass yields for algae require an increase demand for certain nutrients. In conventional crops, yield increases over the past few decades are highly correlated to increased application of nitrogen fertilizers derived from the Haber-Bosch industrial process. This energy-intense nitrogen-fixing process requires substantial inputs of pressure and heat, often tied to natural gas consumption. Thus, the realization of improved yields of algal biomass is likely to be accompanied by an increased demand for Haber-Bosch derived nitrogen. As a result, a portion of the energy obtained thought the production of algal biofuels will need to be diverted back to Haber-Bosch processes in order to sustain the accompanying biomass.

Though the increased demand for industrially produced nitrogen would significantly impact the overall economics of algal biofuel production, nature provides us with examples from current agricultural practices where symbiotic relationships can eliminate the need for externally provided nitrogen. Model relationships between plants such as legumes and the nitrogen-fixing bacteria that are associated with them can provide a blueprint for methods of developing similar relationships with algae. However, while legumes have evolved these relationships in a manner that supports the symbiosis, this is accomplished through compartmentalization, where nitrogen fixation is generally done in a low oxygen environment of the roots, while oxygen-producing photosynthesis is isolated to pigmented tissues. A similar system may be difficult to recreate in large-scale cultures with microalgae.

Studies in our laboratory focus on the aerobic nitrogen-fixing bacterium *Azotobacter vinelandii*, which is a unique and model organism for nitrogen fixation. We have recently made progress toward the development of alternative routes to increase nitrogen production by this bacterium through the application of broad screens, developing terminal nitrogen products and deregulating nitrogen fixation. We have further demonstrated the successful culture of this bacterium with various algae strains by providing a supplemental source of sugar to support the nitrogen fixation in *A. vinelandii*. Many algae are known to produce extracellular sugars and polysaccharides, and our current efforts aim to increase carbohydrate production by the algae during photoautotrophic growth, to establish the mutually beneficial relationship that will support both species.

429370 A Novel Symbiotic Biofilm for Algae Growth and Harvesting

Abstract:

Microalgae has the potential to become a substantial driver in the development of a bio-based economy, and has been considered as a promising source for biofuel and bioenergy production, bio-remediation and other high-value bioactive compounds or combination of above. Large-scale applications of microalgae biotechnology is hampered by the current energy-intensive or non-ecofriendly chemical harvest technology; along with the nutrient and water requirements. In this work, we attempt to develop a lichen-type symbiotic biofilm using algae and fungal cultures and also address the cultivation barriers using a synthetic community which will possibly transform the algae cultivation in the future for various applications. Cultures with algae and fungi could be the ideal strain combination for symbiotic biofilm because these two strains don't compete for the same respiratory gases whereas they complement each other; algae is photoautotropic; and the fungi can utilize wide range of substrates with the history of having wide range of symbiotic functions in the biota.

The lichen-type biofilm formation was studied with various fungal strains and model algae strain *C. vulgaris* and the factors affecting the process including effect of different matrix for attachment, initial glucose concentration, and biomass distribution at different time interval. The possibility of co-existence of *Chlorella vulgaris* with various fungal cultures was tested to identify the best strain combination for high algae harvest efficiency. The effect of different matrix type (woven stainless steel mesh, coiled metal matrix, cotton mesh, and cotton-polypropylene composite mesh) and initial glucose concentration (0-10 g/L) on biofilm formation and algae attachment; and kinetics of the process with respect to the algae-fungi cell distribution and total biomass production was estimated. The algae attachment efficiency of 99.0 % and above was achieved in a polymer-cotton composite matrix with low glucose concentration of 2 g L^{-1} in the growth medium for initiating the fungal growth and agitation intensity of 150 rpm at 27 °C. The ultra-clear water in the flask after removing the biofilm can be used for next batch of algae-fungi culture without any pretreatment or processing with nutrient addition. The existing technologies are economically feasible only to algae production for high-value products as there is a significant cost involved in algae harvest and dewatering. Whereas, the novel microalgae cultivation platform technology is completely different from all the traditional methods and avoids the conventional complex harvesting and processing procedures and propounds the scope for an energy efficient, eco-friendly and cost-effective "lichen" type biofilm, exploring new avenues in the bio-production industry with many potential applications.

432515 Insight into Lipid Biogenesis Using Stable Isotope Tracers Coupled with NMR Spectroscopy and Mass Spectrometry

Abstract:

Algae have great potential for use in biofuel and specialty chemical production due to inherit growth advantages, high triacylglyceride (TAG) content per cell, and the use of renewable energy sources (*i.e.*, sunlight) along with the potential to balance atmospheric carbon levels. However, there is a lack of basic algal research in lipid biogenesis and metabolism that would aid industry and advance the development curve. Algal TAG accumulation is the net result of TAG synthesis minus utilization and typically occurs when cellular cycling is stunted or arrested due to nutrient limitation, environmental stress, or by chemical addition. Here we report on a simplified metabolomics method using ^{1}H and ^{13}C NMR to directly monitor ^{13}C incorporation and recycling in real-time, using live cells, during bicarbonate enhanced TAG accumulation induced by N-deplete culturing. Results indicate rapid carbon incorporation into sucrose (within 15 min) directly after N-depletion, which increased in cellular concentration for 16 hours after which a steady-state concentration was observed. Within four hours, TAG accumulation was observed. After 48 hours 65-75% of the carbon incorporated into fully saturated or mono-unsaturated fatty acids was found to be derived from extracellular dissolved inorganic carbon. Conversely, the majority of poly unsaturated and omega-3 fatty acids that accumulated post N-depletion were recycled presumably from preexisting cellular membranes (biomass synthesized prior to N-depletion). The extent of ^{13}C incorporation into fatty acids, from either extracellular dissolved inorganic carbon, or from recycled intracellular biomass, was confirmed using gas chromatography mass spectroscopy. In conclusion, this work highlights state of the art metabolomics techniques that expand on current metabolomics methods by avoiding cellular extraction while reducing the amount of untargeted metabolite detection and identification. This study answers key lipid biogenesis questions such as the extent to which *de novo* fatty acid synthesis versus recycled fatty acids takes place.

430852 Innovative Algal Process and System for Sustainable Biofuels and Biochemicals Production

Abstract:

In this presentation, results from the research and development of a suite of technologies for mass cultivating microalgae on waste streams and converting the harvested algal biomass into fuels, animal/fish feed and other high value byproducts will be presented. Specifically, the presentation will present and discuss following R&D areas:: 1) development of a fast screening strategy for local bio-prospecting of robust algal stains which adapted well in different types of wastewater resources including municipal wastewater (CMW) and swine manure wastewater (DSM) and the results showed the high lipid productivity (77.5-94.8 mgL-1d-1) and high growth rate (0.43-0.53 d-1) were obtained for the selected most promising strains belonging to Chlorella, Heynigia, Hindakia, Micractinium, and Scenedesmus, etc; 2) development of a low-capital and maintenance cost and highly scalable multi-layer bionic hybrid cultivation systems for year-round production of algal biomass from various waste streams; 3) development of an integrated swine manure-based algal platform for sequential biofuel and omega-3 contained algal biomass production for animal/fish feed uses; 4) development of a novel fungi-assisted bio-flocculation technique for low-cost and highly efficient microalgal cells harvesting; and 5) development of an integrated biorefining concept and system process for water recycling and nutrient reuse, including combing fungi-assisted harvesting system for water recycling and utilization the nutrients of aqueous phase derived from the pyrolysis and hydrothermal liquefaction of the algae biomass for nutrients reuse.

436713 Integrated Computational and Experimental Studies of Microalgal-Based FUEL Production

Abstract:

In recent decades, it has become obvious that the world has been confronted with an energy crisis and the reliance on traditional fossil fuel resources is unsustainable, due to both the irreversible depletion of fossil fuels and the associated greenhouse gas emissions which cause global warming [1]. Production of biofuels from microalgae has the potential to overcome the major drawbacks of the fossil fuels and plant derived fuels, greenhouse gas emissions, energy security and competition for agricultural land [3]. Also, algal oil production is one of the most promising technological eco-innovations [2]. Nevertheless, the technology is in its early stage of development and it is not competitive with the traditional fossil fuels [3].

The current work aims to establish links between algal strains grown in large raceway open ponds and innovative fuel generation technologies in order to achieve positive energy balance and environmental sustainability. Specifically, novel multi-parameter quantification and predictive modelling will be investigated to describe algal growth and lipid accumulation in laboratory-scale batch systems. The purpose of modelling of such closed batch reactors is to identify key parameters, which lead to the microalgae lipid and starch accumulation. Subsequently, the key parameters defined in the batch system through modelling and optimisation studies will be used, in order to further develop novel harvesting techniques as well as oil and starch extraction methods for industrial scale pond use. Ultimately, the validated predictive closed batch system model will be used to construct open raceway pond models for maximization of the biomass production and as well as maximization of oil and starch production in such systems.

A multi-parameter quantification and predictive model is investigated [4] to describe algal growth and lipid and starch accumulation in laboratory-scale batch systems and it has been expanded to cover other growth parameters (temperature, light, pH) [5,6] to control productivity of microalgae cultivation technologies. Experiments have been conducted to analyse the effect of different input parameters and the results have been used to identify key parameters of the laboratory-scale batch systems by using a combination of deterministic and stochastic optimisation methods.

References

1. Hoel, M. & Kverndokk, S. 1996. Depletion of fossil fuels and the impacts of global warming. Resource and Energy Economics, 18, 115-136.
2. Pienkos, P. T. & Darzins, A. 2009. The promise and challenges of microalgal-derived biofuels. Biofuels, Bioproducts and Biorefining, 3, 431-440.
3. Chisti, Y. 2007. Biodiesel from microalgae. Biotechnology Advances, 25, 294-306
4. Economou, C. N., Aggleis, G., Pavlou, S. & Vayenas, D. V. 2011. Modeling of single-cell oil production under nitrogen-limited and substrate inhibition conditions. Biotechnology and Bioengineering, 108, 1049-1055
5. Zhang, X.-W., Chen, F. & Johns, M. R. 1999. Kinetic models for heterotrophic growth of Chlamydomonas reinhardtii in batch and fed-batch cultures. Process Biochemistry, 35, 385-389.
6. Guterman, H., Vonshak, A. & Ben-Yaakov, S. 1990. A macromodel for outdoor algal mass production. Biotechnology and Bioengineering, 35, 809-819.

AUTHOR INDEX

AUTHOR INDEX

AUTHOR INDEX

AIChE
120 Wall Street, FL 23
New York, NY 10005-4020

ISBN 978-1-5108-1861-3

CURRAN ASSOCIATES INC.
proceedings.com